Lukas Metzger

Process Simulation of Technical Precipitation Processes

The Influence of Mixing

Process Simulation of Technical Precipitation Processes

The Influence of Mixing

by
Lukas Metzger

KIT Scientific Publishing

Dissertation, Karlsruher Institut für Technologie
KIT-Fakultät für Chemieingenieurwesen und Verfahrenstechnik

Tag der mündlichen Prüfung: 13. Oktober 2017
Gutachter: Prof. Dr.-Ing. Matthias Kind
Prof. Dr.-Ing. habil. Hermann Nirschl

Impressum

KIT Scientific Publishing

Karlsruher Institut für Technologie (KIT)
KIT Scientific Publishing
Straße am Forum 2
D-76131 Karlsruhe

KIT Scientific Publishing is a registered trademark
of Karlsruhe Institute of Technology.
Reprint using the book cover is not allowed.

www.ksp.kit.edu

Print on Demand 2018 – Gedruckt auf FSC-zertifiziertem Papier

ISBN 978-3-7315-0735-2
DOI 10.5445/KSP/1000076187

Process Simulation of Technical Precipitation Processes

The Influence of Mixing

zur Erlangung des akademischen Grades eines

Doktors der Ingenieurwissenschaften (Dr.-Ing.)

der Fakultät für Chemieingenieurwesen und Verfahrenstechnik
des Karlsruher Instituts für Technologie (KIT)

genehmigte

Dissertation

von

Dipl.-Ing. Lukas Metzger

geboren in Speyer

Tag der mündlichen Prüfung: 13.10.2017
Erster Gutachter: Prof. Dr.-Ing. Matthias Kind
Zweiter Gutachter: Prof. Dr.-Ing. habil. Hermann Nirschl

Kurzfassung

Die Fällungskristallisation zählt zu den wichtigen Feststoffbildungsverfahren in der pharmazeutischen und chemischen Industrie. Sie beschreibt die Bildung von schwerlöslichen, feindispersen Feststoffphasen aus einer Flüssigphasenreaktion von zwei oder mehreren Reaktanten. Die dabei erzeugten Partikel liegen zumeist im Größenbereich einiger Nanometer bis mehrerer Mikrometer. Das Verfahren findet Verwendung in eine Vielzahl von Einsatzfeldern. So werden beispielsweise Batteriematerialien, Pigmente für Farben und Lacke, heterogene Katalysatoren oder auch pharmazeutische Wirkstoffe, um nur einige zu nennen, über diesen Verfahrensschritt synthetisiert. Die charakteristischen Eigenschaften der Fällprodukte, wie Partikelgrößenverteilung, Kristallmorphologie, Habitus oder Porosität hängen stark von der intrinsischen Kinetik des Reaktionssystems sowie der überlagerten Mischungscharakteristik des verwendeten Fällungsapparates ab. Trotz der Etabliertheit des Verfahrens als verfahrenstechnische Grundoperation besteht oftmals ein noch unvollständiges Prozessverständnis. Prozessauslegungen und Scale-up stellen besondere Herausforderungen dar. Diese ergeben sich im Wesentlichen aus dem Spannungsfeld von zum einen sehr schnellen (meist $< 1s$) und damit stark lokalen Primärpartikelbildungsprozessen und zum anderen von langsamen, über die Gesamtprozesszeit ($\sim$min $-$ h) stattfindenden Sekundärmechanismen (reisezeitabhängige Prozesse, Agglomeration und Alterung). Gerade die schnelle Primärpartikelbildung mit sowohl parallel als auch konsekutiv ablaufenden Mikroskalen-Vorgängen, wie Vermischung der Reaktanten, Keimbildung, Wachstum und Aggregation, ergänzt durch die anspruchsvolle Thermodynamik des betrachteten Systems, führt zu schwer vorhersehbaren Zusammenhängen.

Der Fokus dieser Arbeit liegt auf der simulativen Methodenentwicklung für ein tiefergehendes Verständnis komplexer (ionischer) Fällungsprozesse. Dabei werden die genannten Herausforderungen der Multiskaligkeit gezielt angegangen. Die Dissertation gliedert sich in zwei aufeinander aufbauende Teile.

Um zunächst die Komplexität des Systems zu reduzieren werden im ersten Teil der Arbeit nur die Primärprozesse der Fällung am Beispiel des in der Literatur gut untersuchten Modellstoffsystems Bariumsulfat betrachtet. Die Untersuchungen richten sich auf das Verständnis der Interaktion von Vermischung und den Primärprozessen der Partikelbildung auf mikro- und meso-Ebene. Dabei bieten sich Mischdüsen als Benchmark-Apparat an. Neben einer hohen Reproduzierbarkeit der Partikelgrößenverteilungen kann in diesem Apparatetyp der Vermischungseinfluss sensitiv durch die Durchflussraten gesteuert werden. Zudem erlauben Mischdüsen durch ihr schnelles Vermischungsvermögen bei hohen Strömungsgeschwindigkeiten den Ausschluss von Mischungseffekten auf die Partikelbildung. Damit besteht Zugang zur Fundamentalkinetik des Stoffsystems. Methoden der numerischen Strömungssimulation (CFD) ermöglichen in dieser Arbeit Einblick in die sehr schnellen Vermischungsprozesse, die experimentell nur schwer zugänglich sind.

Drei methodische Ansätze auf unterschiedlichem Komplexitätsniveau werden erarbeitet und verglichen. Zum einen wird mittels CFD, direkt gekoppelt mit Populationsbilanzen die Partikelbildung in Mischdüsen zeitlich vollaufgelöst betrachtet. Zweitens wird dieses Vorgehen verglichen mit einem neu erarbeiteten Vergröberungsansatz der im folgenden STAR NM (Spatially and Temporally Averaged Reduced Numeric Measurement) genannt wird. Hierbei wird in einphasigen CFD Simulationen die Zustandsgröße Übersättigung lokal und zeitlich gemittelt und in schnellen eindimensionalen Populationsbilanzrechnungen hinterlegt. In einem dritten methodischen Ansatz werden die Ergebnisse der CFD-Mittelungen aus Methode 2 mit theoretischen Ansätzen aus der klassischen Mischtheorie verglichen und bewertet.

Die Betrachtung dieser drei Methoden ermöglicht neben dem detaillierteren Verständnis der gekoppelt ablaufenden Partikelbildungsmechanismen und der Vorhersage von finalen Partikelgrößenverteilungen auch die Ableitung eines scale-up-fähigen Modells für Mischdüsen. Hier besteht großes Potential, da bis dato Effekte, die durch die Skalierung von Mischdüsen hervorgerufen werden, noch nicht vollständig verstanden sind. Zusätzlich kann gezeigt werden, wie sich die erarbeiteten Methoden in ein Fließschema-Programm integrieren lassen und dabei eine Verallgemeinerung auf andere Stoffsysteme ermöglichen.

Der zweite Teil dieser Arbeit dient dem Methodentransfer auf Rührkesselapparate. Diese stellen den Standardapparat in der industriellen Anwendung dar. Zum einen wird die zuvor erarbeitete Vergröberungsmethodik zur Verknüpfung von mikro- und meso-Skala übertragen und validiert. Zum anderen liegt ein erweiterter Fokus auf den zusätzlichen in Rührkesseln zu berücksichtigenden Skalen (Makro- und Prozesszeitskala). Betrachtet werden Versuche im Semibatch-Betrieb. Die Fällungsprozesse im Rührkessel sind gekennzeichnet durch schnell ablaufende Primärpartikelbildungsprozesse unmittelbar an der Feed-Einleitstelle, überlagert durch eine makroskopische Zirkulation im Apparat. Im Strömungsfeld schon existierende Partikel zirkulieren entlang Ihrer Reisezeit immer wieder durch die Reaktionszone und dienen als sekundäre Keime. Sowohl experimentell als auch simulativ wird gezeigt, dass die Komplexität des Systems gut durch einen Multikompartiment-Ansatz abgebildet werden kann. Experimentell wurde hierzu, unter Berücksichtigung entsprechender Literaturrecherche, ein Mehr-kompartiment-Versuchstand neu aufgebaut. In einem Fließschemamodul wird diese Verschaltung von Kompartimenten simulativ nachgestellt. Experimentelle Untersuchungen können in dieser Arbeit zeigen, dass das konzipierte Ersatzschaltbild in der Lage ist Referenzversuche im Rührkessel abzubilden. Ergänzend wird dies in der Fließschemasimulation bestätigt.

Zusammenfassend werden im Rahmen dieser Arbeit zwei skalenverknüpfende Methoden (Vergröberungs- und Multikompartiment-Ansatz) erarbeitet und validiert. Diese ermöglichen es die bestehende zeitliche und lokale Diskrepanz zwischen mikro-, meso- und makro-Skala zu schließen und damit das bestehende Prozessverständnis signifikant zu verbessern.

Abstract

Precipitation crystallization is one important unit operation of solid formation in the pharmaceutical, chemical and process industry. It depicts the formation of hardly-soluble, fine-disperse solid phases from a liquid phase reaction of two or more components. Commonly, the particles processed possess the size range of some nanometers up to several micrometers. The application of precipitation is a highly diverse field. The processing of battery materials, of pigments for paints and coatings, of heterogeneous catalysts and precursors or for instance the synthesis of pharmaceutical active ingredients, to name only a few, are common applications. Characteristic attributes of the final products, such as particle size distribution, morphology, habitus and porosity, strongly depend on the intrinsic kinetics of the reactant system and on the mixing characteristics of the particular apparatus. Despite the matter that precipitation is an established unit operation, process understanding is yet insufficient. Process design and scale-up represent a particular challenge. This mainly results from the interplay of very fast and therefore, highly spatial primary processes of particle formation (usually < 1s) and from the slow secondary mechanisms that take place along the overall process time (residence time dominated processes, agglomeration, aging; from min to h). Especially fast primary processes, incorporating parallel and consecutive occurring microscale-processes such as the mixing of reactants, nucleation, growth and aggregation which are complemented by demanding thermodynamics of the system regarded, leads to coherences which are hard to predict.

This work focuses on the method development using certain simulative tools to get a deeper insight into such complex (ionic) precipitation processes. Thereby, previously mentioned multi-scale challenges are aimed to be tackled. The contribution is structured into two parts that build on one another.

In order to reduce the complexity of the system the first part of this work only considers the primary process of precipitation using the model system barium sulfate which is well investigated in literature. The investigations are focused on the understanding of molecular, micro- and meso-scale interactions. Thereby, confined impinging jet mixers (CIJMs) are appropriate apparatuses. Besides a high reproducibility of particle size distributions, this type of apparatus allows a sensitive control of mixing by varying the flow rates. Moreover, CIJMs enable, due to their high mixing capability at high flow velocities to suppress mixing effects influencing the particle formation. Thus, access to the fundamental kinetic of the precipitating system is given. In this work, computational fluid dynamics (CFD) enable insight into the fast mixing processes which is hard to get experimentally. Three methodological approaches are developed and compared, differing in their complexity. Firstly, by using CFD which is coupled with population balances, the particle formation in CIJMs is temporally resolved. Secondly, this procedure is compared to a new developed coarse-graining approach which is called STAR NM (Spatially and Temporally Averaged Reduced Numeric Measurement). Hereby, single phase CFD simulations are used to deposit spatially and temporally averaged state variables such as the supersaturation into fast one-dimensional population balance calculations. Thirdly, the results from method 2 are compared to theoretical approaches form the classical mixing theory. Besides the acquirement of a detailed process understanding and the ability to forecast final particle size distributions the consideration of these three methods leads to the derivation of scale-up model for CIJMs. Hereby, an interesting potential exists, since at present, scaling of such mixers is not fully understood. Additionally it can be shown that the methods worked out can be integrated into a flowsheet tool where a further generalization on other material systems can be done.

The second part of this work serves for the method transfer on stirred tank reactors (ST) which is the most commonly used apparatus in industry. On the one hand the coarse-graining method from section 1 is transferred and validated. On the other hand an extended focus is set on additional scales (macro- and process scale) being present in STs. An operation is semi-batch mode is regarded. The procedures in such apparatuses are characterized by fast primary processes straight after the feed inlet which are superposed by a macroscopic circulation in the apparatus. Particles that already exist in the flow field travel along their residence time over and over again through the reaction zone and serve for secondary nuclei. Experimentally and simulatively it is shown that the complexity of the system can be abstracted well by a multi-compartment approach. In consideration of the literature an experimental multi-compartment setup is newly built. Additionally, in a flowsheet module this circuitry is adjusted simulatively. Experimental investigations in this work are able to show that the conceptualized equivalent circuit is able to reproduce reference experiments in a ST. Additionally this is confirmed by the flow sheet simulation.

In comparison, two scale-comprehensive methods (coarse-graining and multi-compartment approach) are developed and validated in this work. They enable to work as a closure for the existing temporal and local discrepancy between micro-, meso- and macro-scale and moreover, to improve the existing process understanding.

Content

Kurzfassung i

Abstract v

Vorwort xiii

Preface xv

List of Symbols xvii

Latin letters xvii
Greek letters xix
Indices xx
Dimensionless numbers xxi
Constants xxii
Abbreviations xxii

1 Motivation 1

1.1 Aim and scope of this work 2

2 Methods and fundamentals 5

2.1 General population balance equation 5
2.1.1 Algorithms applied – High resolution finite volume method (HRFVM) 6
2.1.2 Algorithms applied – Direct quadrature method of moments (DQMOM) 8
2.2 Thermodynamic fundamentals 10
2.2.1 Supersaturation 13
2.2.2 Free lattice ion ratio 14
2.2.3 Activity coefficient approaches 14
2.3 Precipitation 19
2.3.1 Nucleation 19
2.3.2 Growth 23
2.4 Simulation assumptions using $BaSO_4$ as a model system 27

3 The influence of mixing –Method development using the confined impinging jet mixer as a benchmark apparatus 33
3.1 Experimental Setup 34
3.1.1 Laser-spectroscopic flow field investigations...................... 35
3.1.2 Precipitation experiments 36
3.1.3 Characterization of precipitated particles.......................... 37
3.2 Numerical setup and turbulence 38
3.3 Fluid mechanics and precipitation in CIJMs.................................. 41
3.3.1 Experimental section... 42
3.3.2 Numerical section and comparison to experiments 51
3.4 Method development – coupling mixing and particle formation... 60
3.4.1 The influence of "sub-grid" mixing within CFD 61
3.4.2 Coupled computational fluid dynamics - population balance equation (CFD-PBE) approach 67
3.4.3 STAR NM – spatially and temporally averaged reduced numeric measurement... 75
3.4.4 Classical mixing theory 94

4 CIJM – Flow sheet simulation and approach potential...................... 111
4.1 Flow sheet simulation.. 111
4.2 The influence of mixing using the limiting time scale approach .. 112
4.3 The influence of supersaturation 114
4.4 Scale up of CIJMs 115
4.5 Transferability to other material systems 117
4.6 Interconnections – consecutive combined CIJMs........................ 118
4.7 Computational time... 119

5 Method transfer – stirred-tank reactors 121
5.1 Aim and scope 121
5.2 State-of-the-art – multicompartment approaches....................... 123
5.3 Simulation strategy.. 125
5.4 Experimental setup.. 127
5.4.1 Design of the holdup vessel R_2 128

5.4.2 Design of a stirred-tank reaction compartment – experimental concept A ... 132
5.4.3 Design of a jet in crossflow reaction compartment – experimental concept B ... 133
5.5 Numerical section – estimation of process-dominating parameters by STAR NM ... 136
5.5.1 Numerical setup ... 136
5.5.2 Numerical results ... 138
5.6 Results – experimental investigations ... 148
5.6.1 Experimental reference – stirred-tank investigations ... 148
5.6.2 Experimental simulation – a compartment approach ... 150
5.7 Results – flowsheet modeling ... 152
5.7.1 Variation 1 – two-compartment mixer holdup-assembly 153
5.7.2 Variation 2 – two-compartment jet in crossflow ... 156

6 Conclusion and Outlook ... 163
6.1 Method development – confined impinging jet mixer ... 164
6.2 Method transfer – stirred-tank reactor ... 166

7 References ... 169

8 Appendix ... 179
8.1 Methods and Materials ... 179
8.1.1 Reference potentials ... 179
8.1.2 Material data used for $BaSO_4$ ... 179
8.2 Confined Impinging Jet Mixers ... 181
8.2.1 Refractive indices correction for LIF and PIV ... 181
8.2.2 Flow regimes in CIJMs – CFD findings ... 185
8.2.3 Additional STAR NM results ... 186
8.2.4 Cavitation phenomena in CIJMs ... 192
8.2.5 Introducing a density and rheology model for BaSO4 suitable for the application within CFD ... 194
8.3 Dyssol – a simulation framework system based on C++ ... 198
8.4 Stirred-tank reactors ... 199

8.4.1 Influence of the feed pipe diameter 199
8.4.2 Reaction compartments R1 199
8.4.3 Turbulence model and sliding mesh procedure 200
8.5 Mathematical annotations 201
8.5.1 DQMOM algorithm for CFD 201
8.5.2 Product difference algorithm 205

List of publications 207
Publications 207
Conference contributions 208
Student theses conducted in conjunction with this thesis 209

Vorwort

Die Inhalte der vorliegenden Arbeit entstanden während meiner Zeit als wissenschaftlicher Angestellter und Doktorand am Institut für Thermische Verfahrenstechnik (TVT) am Karlsruher Institut für Technologie (KIT) zwischen November 2011 und November 2016.

Zur Entstehung und zum Gelingen dieser Arbeit trugen viele Menschen bei, denen an dieser Stelle gedankt werden soll.

Besonderer Dank gilt meinem Doktorvater, Herrn Prof. Dr.-Ing. Matthias Kind für seine Unterstützung und die Förderung meiner Arbeit, für die wissenschaftlichen Freiheiten und das mir entgegengebrachte Vertrauen.

Herrn Prof. Dr.-Ing. habil. Hermann Nirschl danke ich für die freundliche Übernahme des Korreferates und für das meiner Arbeit entgegengebrachte Interesse.

Bei meinen Freunden und Kollegen möchte ich mich ganz herzlich für das angenehme Arbeitsklima und die gute Zusammenarbeit bedanken. Danke für die vielen fröhlichen Stunden innerhalb und außerhalb des Instituts. Ganz besonderer Dank gilt meiner Bürokollegin Esther Laryea und meinen Kollegen Ricco Kügler, Philipp Lau und Zhen Li für die vielen fachlichen und konstruktiven Diskussionen.

Einen wichtigen Beitrag zum Gelingen dieser Arbeit liefern die Studentinnen und Studenten, die unter meiner Anleitung ihre Studien-, Bachelor- und Masterarbeit angefertigt haben. Dies sind Jan Udri, Pascal Hirschel, Klaus Müller, Kim Nguyen, David Guse, Marvin Winkler, Matthias Kunz, Sonja Seiler, Marijan Mohr, Tobias Nachtrab, Michael Zeller, Jan Sauer, sowie meine wissenschaftlichen Hilfskräfte Markus Busch und Stephan Höcker. Ein herzliches Dankeschön für ihr Engagement und die vielen fachlichen aber auch persönlich bereichernden Gespräche.

Im Weiteren danke ich Nicole Feger, Rebekka Golling und Tamara Cataldo für die administrative Unterstützung im Institutssekretariat. Ich danke Sebastian Schwarz für seine Konstruktionen sowie den Herren aus der Werkstatt, Michael Wachter, Steffen Haury und Stefan Fink, die mich tatkräftig beim Aufbau des Kompartment-Versuchstands unterstützt haben.

Danke auch an Volker Zibat vom Laboratorium für Elektronenmikroskopie (LEM) am KIT für die in dieser Arbeit enthaltenen REM Aufnahmen.

Der Deutschen Forschungsgemeinschaft (DFG) gilt mein Dank für die finanzielle Förderung meiner Arbeit im Rahmen des Schwerpunktprogramms 1679 „Dynamische Simulation vernetzter Feststoffprozesse – DynSim-FP“. Spezieller Dank gilt Daniele Marchisio (Politecnico di Torino) für seine Unterstützung beim Aufbau der DQMOM-IEM Simulation.

Zu guter Letzt möchte ich mich besonders bei meinen Eltern bedanken, die mich über die gesamte Zeit meiner Promotion unterstützt haben und mir immer mit Rat und Tat zur Seite standen. Ebenso danke ich meiner Freundin für ihren Rückhalt und ihre Liebe.

Karlsruhe, November 2016

Lukas Metzger

Preface

The content of this work has been evolved in my time as a scientific employee and PhD student at the Institute of Thermal Process Engineering (TVT), Karlsruhe Institute of Technology, from November 2011 to November 2016.

I would like to take this opportunity to thank many people who contributed to the success of this work.

I gratefully thank my supervisor Prof. Dr.-Ing. Matthias Kind for his support, for the encouragement and the scientific freedom as well as for the steady confidence throughout this work.

Furthermore, I would like to thank Prof. Dr.-Ing. habil. Hermann Nirschl for providing the second review of this thesis.

Warm thanks to my friends and colleagues for the convivial working atmosphere and the nice cooperation. Thanks for all the good times inside and outside of the institutional framework. Special thanks to my office colleague Esther Laryea and to Ricco Kügler, Philipp Lau and Zhen Li for many scientific and constructive discussions.

An important contribution to this work was made by my students with their Bachelor and Master theses: Jan Udri, Pascal Hirschel, Klaus Müller, Kim Nguyen, David Guse, Marvin Winkler, Matthias Kunz, Sonja Seiler, Marijan Mohr, Tobias Nachtrab, Michael Zeller and Jan Sauer, as well as my student research assistants Markus Busch and Stephan Höcker. Special thanks for their commitment and the many technically and privately rewarding conversations.

My gratitude goes to Nicole Feger, Rebekka Golling and Tamara Cataldo for the helpful administrative support in the TVT head office. Thanks to Sebastian Schwarz for his constructions and to the TVT workshop crew: Michael Wachter, Steffen Haury and Stefan Fink for the energetic support while constructing the compartment laboratory plant.

Additional thanks to Volker Zibat from the laboratory of electron microscopy (LEM) at KIT for the scanning electron microscope images included in this work.

The financial support of the DFG (Deutsche Forschungsgemeinschaft) within the priority program SPP 1679 "Dynamic flowsheet simulation of interconnected solids processes" is gratefully acknowledged. Special thanks to Daniele Marchisio (Politecnico di Torino) for his helpful support in setting up the DQMOM-IEM model.

Finally, I would like to thank particularly my parents for the continuous support throughout my PhD thesis and their helpful assistance and advice. Furthermore, I thank my girlfriend for her backing and her love.

Karlsruhe, November 2016

Lukas Metzger

List of Symbols

Latin letters

a_i	activity of component i (-)
$a_1, a_2, a, A_1,$	empirical constants $(var.)$
$a_{eff,i}$	effective diameter of ion i (Å)
A_c	constant in eq. 3.42 (-)
A_{DH}	Debye-Hückel constant (-)
A	surface (m^2)
b_1, b_2, b_3	dimensions of a stirred tank (m)
B	birth rate (nucleation) $(m^{-3}s^{-1})$
B_i	Jones Dole parameter of component i (-)
$\mathcal{B}$	matrix
b, b_0, B_1, B_2	empirical constants (-)
$\tilde{c}_i$	molar concentration of component i (mol/l)
C_{Des}, C_μ	constants $(var.)$
$C_{\xi,max}$	empirical constants in eq. 3.21(-)
C_ξ	mixing constant in eq. 3.19(-)
d_h	hydraulic diameter (m)
d_2	impeller diameter (m)
d_i	diameter i (m)
d_{ij}	characteristic particle diameter $\equiv \mathcal{M}_i/\mathcal{M}_j$ (m)
D_{AB}^{α}	diffusion coefficient of A in B in phase α (m^2/s)
D_{ax}	axial dispersion coefficient (m^2/s)
D_i	inner diameter of reactor i (m)
E	engulfment rate coefficient $(1/s)$
f	vortex shedding frequency $(1/s)$
f_t	temporal safety factor (-)
f_i	factor (-)
F_2, f_d	blending functions (-)
F_{i+}, F_{j-}	cation and anion contribution parameters (-)
$\mathcal{F}$	flux $(1/m^2)$
G^E	Excess Gibbs free energy (J)
h_1, h_2	dimensions of a stirred tank (m)
h_i^L	relative liquid level in compartment i (-)
H_i	height of reactor i $(H_{i,max} = D_i)$ (m)
$I_{\tilde{c}}$	molar ionic strength (mol/m^3)
$I_{\tilde{m}}$	molal ionic strength $(mol/kg$ solvent)
I_S	Intensity of segregation (-)

k	turbulent kinetic energy (m^2/s^2)
k	wave number of a vortex $(1/m)$
k_V, k_A	volumetric and surface shape factor (-)
K_0	empirical constant (-)
K	bulk entrainment coefficient $= f_{circ}/f_{prim}$
K_{IP}	equilibrium constant of the ion pair complex (mol^ν/l^ν)
K_{sp}	solubility product (mol^ν/l^ν)
l_i	weighted abscissa $= w_i\mathcal{L}_i$ (m)
L_i	length of i (m)
L^P	particle size (m)
$L^P_{50,i}$	median value of the particle size distribution q_i (m)
$\mathcal{L}_i$	abscissa i (m)
$\mathcal{L}_p$	noise level (dB)
m	mass (kg)
$\dot{m}$	mass flow (kg/s)
m_0, m_1	empirical parameters (-)
$\widetilde{m}_i$	molality of component i $(mol/kg$ solvent)
$\widetilde{M}_i$	molar mass of component i (kg/mol)
$\dot{M}_i$	mass flow i (kg/s)
$\mathcal{M}_k$	k^{th} moment (m^k)
n_1	empirical constants $(var.)$
n	particle number (-)
n_{ri}	refractive index of substance i (-)
n_{crit}	number concentration of critical clusters (mol/m^3)
n_i	Molar concentration of component i (mol/m^3)
n_{R_i}	stirrer speed in reactor i (s)
N	number of nodes (-)
N_i	moles of component i (mol)
p_i	probability i (-)
p	pressure (Pa)
P_k, P_ω	production terms of k and ω $(var.)$
q_0	number density $(1/m)$
q_3	volume density $(1/m)$
r_i	reaction rate $(1/s)$
r	radius (m)
R_c	concentration-based free lattice ion ratio (-)
R_i	reactor number i (-)
s_i	weighted probability i $(= p_i\xi_i)$ (-)
$S_{i,k}$	source term of mechanism i of moment k $(var.)$
S	entropy (J/K)

S_a	activity-based supersaturation (-)
t	time (s)
T	temperature (K)
$\vec{u}$	velocity vector (m/s)
$\bar{\vec{u}}$	Re-averaged velocity vector (m/s)
u	velocity (m/s)
U	internal energy (J)
V_i	volume i (m^3)
V_m	molar volume $= \widetilde{M}^\alpha/(\rho^\alpha N_a)$ (m^3/mol)
$\tilde{V}_m$	volume of one mole $= \widetilde{M}^\alpha/\rho^\alpha$ (m^3)
w_i	weight i (-)
$\tilde{x}_i$	molar mass fraction of component i (-)
x_i	mass fraction of component i (-)
x, y, z	spatial coordinates (m)
y_i	normalized radius of compartment i (-)
Y_K, Y_ω	dissipation terms of k and ω (var.)
z_{i+}, z_{i-}	charge of a cation and an anion i (-)
z_{mix}	mixing length coordinate (m)
z_1, z_2	number of internals in a stirred tank (-)
Z_s	scale-up factor (-)
Z	Zeldovich factor (-)

Greek letters

α_i	volume fraction i (-)
$\alpha_{i,u}$	volume fraction i of mesomixed fluid (-)
β_c	collision rate $(1/s)$
β	mass transfer coefficient (m/s)
γ_i	activity coefficient of component i (-)
$\gamma^{\alpha\beta}$	interfacial energy between phase α and β (N/m^2)
$\dot{\gamma}$	shear rate $(1/s)$
δ	Dirac delta function $(1/m)$
$\delta_{i,j}$	Kronecker delta (-)
Δ_{max}	maximal grid spacing (m)
Δp	pressure drop (Pa)
ε	energy dissipation (W/kg)
$\bar{\varepsilon}$	mean energy dissipation (W/kg)
η_t	turbulent viscosity $(Pa \cdot s)$
η_{mix}	laminar mixing efficiency (-)
η^α	dynamic viscosity of phase α $(Pa \cdot s)$
η_r	relative dynamic viscosity $= \eta/\eta_{H2O}$ (-)

θ, ψ	vectors in eq. 8.23
λ	wavelength (m)
λ_B	Batchelor length scale (m)
λ_K	Kolmogorov length scale (m)
Λ_c, λ_{OC}	initial scale (m)
λ_{ij}	binary interaction parameter (-)
μ_i	chemical potential of component i (J/mol)
μ_{ijk}	ternary interaction parameter (-)
ν	moles of reactants necessary to build 1 mole of solute (-)
ν_{kin}	kinematic viscosity (m^2/s)
ν_+, ν_-	stoichiometric coefficients (-)
ξ	pressure loss coefficient (-)
ξ_i	mixture mass fraction (-)
ρ^i	density of phase i (kg/m^3)
$\sigma^{\alpha\beta}$	interfacial tension (N/m)
σ	variance (-)
σ_0	variance of the mean (-)
τ_i	characteristic time i (s)
τ_R	residence time (s)
τ_{50}	median value of the residence time (s)
ϕ	random scalar $(var.)$
ϕ_e	electrical potential (V)
$\phi_{\mathcal{F}}$	flux limiter (-)
ϑ	temperature $(°C)$
ω	char. frequency of energy-dissipating vortices $(1/s)$

Indices

Subscript characters

(aq)	solvated	h	hydrated
$\bar{\alpha}$	spatial average of a variable α	hom	homogeneous
$\langle\alpha\rangle$	temporal average of variable α	jet	related to the jet zone
$'$	fluctuation	L	lower-order
0	pure state	LES	large eddy simulation
∞	ideally dilute state	$meso$	mesoscale
A	surface	$micro$	microscale
A, M, C, A'	mixing regions	mix	mixing
$CIJM$	confined impinging jet mixer	N	standardized
c	characteristic	nom	nominal value
$circ$	circulation	Nuc	nucleation
$crit$	critical	$prim$	primary

$DDES$	delayed detached eddy simulation	$RANS$	Reynolds-averaged Navier-Stokes
$depl$	depletion	ref	reference
ent	entrained	sol	solvent
$free$	not bound in a complex	ST	stirred tank
E	Excess	$STAR\ NM$	gained by STAR NM
f	with additional safety factor	t	turbulent
G	growth	tot	total
H	higher-order	V	volumetric

Superscript characters

$*$	saturated
$\sim$	molar
'	local
-	average
α, β	phase
L	liquid
P	particular
r	rational
S	solid
V	vapor

Dimensionless numbers

CFL	Courant-Friedrichs-Lewy number	$\mathrm{CFL} = \frac{u \cdot \Delta t}{\Delta x}$
Bo	Bodenstein number	$\mathrm{Bo} = \frac{u \cdot L_c}{D_{ax}}$
$\mathrm{Da_t}$	turbulent Damkoehler number	$\mathrm{Da_t} = \frac{\tau_{mix}}{\tau_{depl}}$
$\mathrm{M_R}$	momentum ratio	$\mathrm{M_R} = \frac{\rho_1^L u_1}{\rho_2^L u_2}$
M	M-factor	$M = \frac{\tau_{meso}}{\tau_{micro}}$
Ne	Newton number	$Ne = \frac{P}{\rho n^3 d_2^5}$
Pe	Peclet number	$\mathrm{Re} \cdot \mathrm{Sc}$
$\mathrm{Re_R}$	rotational Reynolds number	$\mathrm{Re_R} = \frac{n \cdot d^2}{\nu_{\mathrm{kin}}}$

Re	Reynolds number	$\mathrm{Re} = \frac{u \cdot L_c}{\nu_{\mathrm{kin}}}$
Sc	Schmidt number	$\mathrm{Sc} = \frac{\nu_{kin}}{D_{AB}^L}$
Sh	Sherwood number	$\mathrm{Sh} = \frac{\beta \cdot L_c}{D_{AB}^L}$
St	Stokes number	$\mathrm{St} = \frac{t^P}{t^L}$
Sr	Strouhal number	$\mathrm{Sr} = \frac{f \cdot L_c}{u}$
ϕ_ε	dimensionless energy dissipation rate	$\phi_\varepsilon = \frac{\varepsilon'}{\bar{\varepsilon}}$

Constants

R	universal gas constant	$8.314\, J/(mol \cdot K)$
N_{a}	Avogadro constant	$6.02214 \cdot 10^{23}\, mol^{-1}$
k_{B}	Boltzmann constant	$1.380649 \cdot 10^{-23}\, J/K$
$\mathcal{F}$	Faraday constant	$96485.33\, C/mol$

Abbreviations

2C	two-compartment approach
2D	two-dimensional
CFD	computational fluid dynamics
CIJM	confined impinging jet mixer
CNT	classical nucleation theory
CSTR	continuous stirred-tank reactor
DDES	delayed detached eddy simulation
DES	detached eddy simulation
DQMOM	direct quadrature method of moments
EED	engulfment-deformation-diffusion model
E-model	engulfment model
GMA	global mixing approach
HRFVM	high resolution finite volume method
IEM	Interaction by exchange with the mean
JiC	Jet in Crossflow arrangement
LES	large eddy simulation
LIF	laser-induced fluorescence

LTSA	limiting time scale approach
ODE	ordinary differential equation
PB	population balance
PBE	population balance equation
PDF	probability density function
PIV	particle image velocimetry
PSD	particle size distribution
STAR NM	spatially and temporally reduced numeric measurement
QMOM	quadrature method of moments
RNG	Re-normalization group method
SST	shear stress transport
ST	stirred tank

1 Motivation

Precipitation crystallization is an important unit operation for the formation of crystalline and amorphous solids. Many applications are settled not only in medicine, biology and analytical chemistry, but also in the chemical and process industry, for example, regarding paints, coatings or catalysts. Thereby, typical particle sizes of precipitation products range between a few nanometers and several micrometers. Supersaturation, as the driving force for the primary processes of particle synthesis, such as nucleation and growth, is a transient and local quantity that can, for instance, be created by a chemical reaction [1]. The driving force generated is usually significantly higher than the ones known for common crystallization processes. Hence, time scales for precipitation are very short and range from milliseconds to seconds. Particle formation during such fast processes is frequently coupled directly to the flow field and the predominant mixing level of ions involved. Consequently, characteristic attributes of the final product, such as particle size distribution (PSD), form, structure and porosity, depend strongly on the intrinsic kinetics of the reactant system and on the mixing characteristics of the particular apparatus.

On the contrary, macroscopic flow phenomena and overall process times can range from minutes to hours. Thus, the scales a process simulation for such precipitation processes must span over several decades in time and space. Figure 1.1 gives a schematic idea regarding this challenging range. Scale-comprehensive simulations and a reasonable coarse-graining are important tools when setting up such a process module. Despite many efforts made in this topic, predictive simulation of the evolving particle collective is still an open field which includes numerous open questions. Crucial aspects that originate directly from this multidimensional nature, such as the scale-up of precipitation apparatuses and its consequences on the particle formation, are still not well understood. Moreover, working concepts for the prediction of precipitation processes are generally thin on the ground.

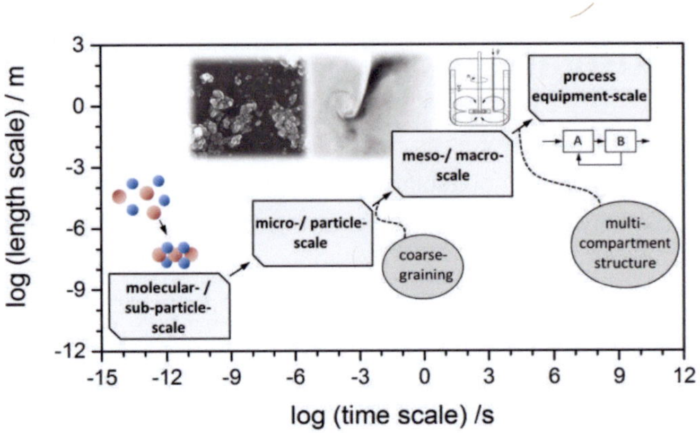

Figure 1.1: Scheme of relevant scales that should be covered for the process simulation of precipitation processes. Their links advanced in this thesis are colored in grey.

1.1 Aim and scope of this work

This work tries to tackle existing deficits in the simulation of precipitation processes. The basic working hypothesis that underlies this thesis can be formulated as follows.

> *Flow sheet simulation can handle complex parallel and consecutively occurring precipitation phenomena, such as mixing, nucleation, growth and recirculation. Thereby, the balancing act of strongly differing timescales, such as milliseconds for the solid formation and seconds to minutes for the overall process time, can be resolved by two important components. On the one hand, coarse-graining of detailed CFD simulations acts as a key element. On the other hand, multicompartment interconnections offer a further important degree of freedom.*

The work aims at validating this hypothesis. To this end it is structured basically into two parts. In the first, the method development is progressed using a confined impinging jet mixer (CIJM) as a benchmark apparatus. This apparatus allows one to study the coupling of molecular scale, microscale and mesoscale. Thereby, the macro and process scale play either no or only a subordinate role. On the one hand, a detailed insight into the coupling of particle formation and mixing by fully resolved CFD-population balance equation (PBE) methods is given. On the other hand, new ideas concerning a reasonable coarse-graining are presented. These coarse-graining data are, furthermore, advanced into a more general concept by using the classical mixing theory which then allows the proposal of a scale-up hypothesis.

In the second part of this thesis, a method transfer is carried out to stirred-tank (ST) reactors in semi-batch operation mode. Thus, not only is the coarse-graining methodology further refined and proven regarding its reliability, but also additional scales (macro and process scales) are added (see Figure 1.1).

The investigations focus on the precipitation of low soluble inorganic materials out of aqueous solution, which is a broad and common class of precipitate. Barium sulfate is used as a highly mixing-sensitive model system which has been intensively studied over the last few decades. Some remarkable work has been published by Gradl et al. [2], Kügler et al. [3], Kucher and Kind [4], Schwarzer [5], Steyer [6] and Vicum and Mazzotti [7]. It is assumed that basic trends found for $BaSO_4$ are transferable to other systems of low solubility. First transfers supporting this assumption will be presented in section 3.

The precipitation processes looked at here are influenced mainly by a set of relevant process parameters which can be classified into a material and an apparatus "function". These parameters span the simulation space that must be covered and which will be considered (partially) in this work.

Material function (depending on the reactants)

- *nominal supersaturation level* $S_{a,\mathrm{nom}}$
- *stoichiometric ratio and the ionic strength*
- *surfactants*
- *temperature and viscosity*

Apparatus function (depending on the apparatus)

- *mixing conditions*

The influence especially of mixing (apparatus function), which is a scale-comprehensive phenomenon, increases the complexity of process simulation strongly. Therefore, special attention is given to this topic and to its superposition to the material function.

2 Methods and fundamentals

This chapter introduces methods and fundamentals of precipitation crystallization which play a crucial role throughout this work. The chapter tries to focus on theoretical approaches that are applied effectively later in chapters 3, 4 and 4. Recommended literature is given in the specific sections for further reading on topics that are not addressed.

Firstly, the PBE, considered as one central element of this work to predict the particle formation during precipitation, is presented. This work concentrates solely on formation processes which are driven by a difference in the chemical potential (primary processes). Hereby, particle number concentration-driven secondary processes, such as agglomeration, breakage or aging, are not considered. Experimental evidence that legitimate such considerations are given in section 2.4. Suitable numerical algorithms to solve such PBEs which are nonlinear integrodifferential equations that do not have an analytical solution are presented. Secondly, the framework of fundamentals and theoretical models, thermodynamics and kinetics which are applied to describe the precipitation processes suitable are introduced. The influence of mixing and its consideration within particle formation modeling are highlighted separately in the appropriate chapters regarding the specific apparatus CIJMs (section 3) and ST reactors (section 4).

2.1 General population balance equation

The PBEs represent a central element, predicting PSDs of a collective fluctuating in time mathematically. Equation 2.1 represents the PBE according to [8] for the crystal phase, considering nucleation, respectively, birth B and size-dependent growth G for an internal coordinate (L^P). In this work, one must distinguish between a one-dimensional treatment, tracking only the internal variable particle size L^P in time (see eq. 2.1), and an approach which is suitable for the CFD application (see eq. 2.2). Thereby, a convection term considering the flow field with the external coordinates (x, y, z) for the three-dimensional (3D) case is included.

$$\frac{\partial n(L^P,t)}{\partial t}+\frac{\partial\left(G\big(S_a(t)\big)\cdot n(L^P,t)\right)}{\partial L^P}=B\big(S_a(t)\big)\cdot\delta\left(L^P-L^P_{\text{crit}}\big(S_a(t)\big)\right) \qquad 2.1$$

$$\frac{\partial n(L^P,t)}{\partial t}+\nabla\cdot\left(\vec{u}\cdot n(L^P,t)\right)+\frac{\partial\left(G\big(S_a(t)\big)\cdot n(L^P,t)\right)}{\partial L^P} =B\big(S_a(t)\big)\cdot\delta\left(L^P-L^P_{\text{crit}}\big(S_a(t)\big)\right) \qquad 2.2$$

Analytical solutions of such PBEs do not generally exist except for a limited number of simpler problems, for example, solely for constant growth. By applying numerical methods to solve the PBEs, the problem of numerical diffusion and dispersion, which are artificial, mathematically owed, error terms, arises [9]. An example of severe defective numeric solutions is given in section 2.1.1. Literature provides a multiplicity of methods that can overcome such problems. An overview of suitable methods and applications can be found in Ramkrishna and Singh [10] and Li [11]. This work focuses on two approaches. On the one hand, an advanced finite volume method (HRFVM) algorithm, which showed a good performance in recent works [11], is used for a discrete, class-based solution of the PBE shown in eq. 2.1.

In terms of CFD applications, such an approach is computational highly cost-intensive. This is owed to the fact that for every class i with a specific internal coordinate value (e.g. particle size L^{P_i}), a specific scalar equation in the 3D flow field must be solved for every grid cell. Therefore, a moment based method (Direct Quadrature Method of Moments-DQMOM) is used, which only conserves the first moments, such as number, length, surface and volume, instead of the full PSD, to overcome this immense computational load. This procedure reduces the set of equations to be solved in CFD to a few additional scalar equations that will be explained in section 2.1.2.

2.1.1 Algorithms applied – High resolution finite volume method (HRFVM)

The HRFV method represents a combination of first and second order discretization which improves the accuracy of the numerical solution significantly.

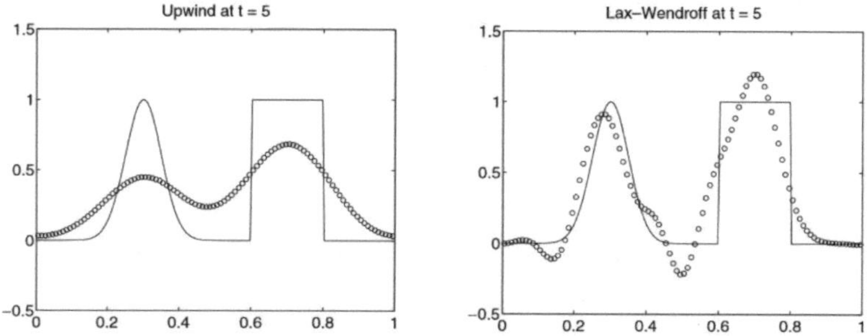

Figure 2.1: Influence of the discretization method: Temporal evolution of two different, constantly growing signals calculated with periodic boundary conditions. One revolution belongs to one time step of length 1, cited from LeVeque [9]; left: First-order upwind, right: Second-order scheme.

First-order methods (upwind scheme) tend to produce numerical diffusion, where the solution of a growing function gets smeared and poor accuracy is reached (see Figure 2.1, left). By contrast, second-order solutions, such as the Lax-Wendroff method (central differencing scheme), lead to numerical dispersion which has the appearance of nonphysical oscillations, especially at discontinuities (see Figure 2.1, right) [12, 9, 13].

The basic idea of the HRFVM is the gainful combination of both fluxes. $\mathcal{F}_L$ in eq. 2.3 represents the lower-order flux (e.g. first-order upwind) and $\mathcal{F}_H$ is the higher-order flux (e.g. second-order, Lax-Wendroff). $\phi_{\mathcal{F}}$ is the blending function, called the flux limiter [9].

$$\mathcal{F}_{i+1} = \mathcal{F}_L + \phi_{\mathcal{F}}[\mathcal{F}_H - \mathcal{F}_L] \qquad 2.3$$

The idea is to combine the advantages of both discretization methods, such as the robustness and the well-capturing of discontinuities of first-order methods, as well as the good accuracy of the higher-order discretization.

Several flux limiters $\phi_{\mathcal{F}}$ can be found in literature which blend higher and lower-order fluxes. Some simple boundary values can be seen easily. The first-order upwind method is applied when $\phi_{\mathcal{F}} = 0$. The second-order scheme is provided when $\phi_{\mathcal{F}} = 1$.

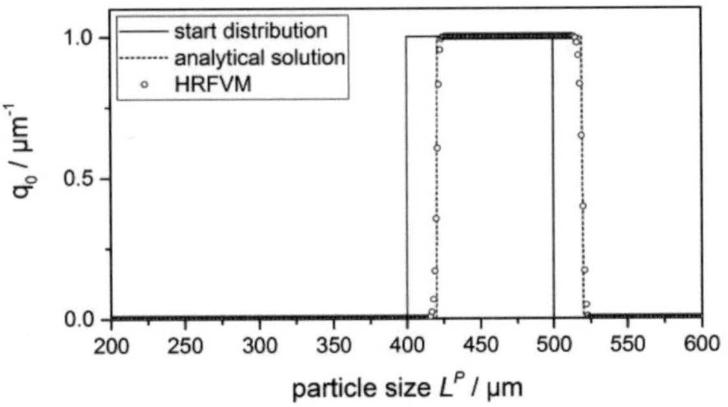

Figure 2.2: Constant growth of a rectangular profile function using the HRFVM algorithm with the van Leer limiter; data are taken from Li [11].

Some more sophisticated limiters can be found in LeVeque [9]. Gunawan [12] recommends the van Leer limiter, which delivered good performances and is used in this work. LeVeque [9] is recommended for more details.

The Courant-Friedrichs-Lewy (CFL number) stability criterion on an equidistant grid is used to calculate the time step size adaptively for the solver maximal allowed at every time point for the numeric solution of eq. 2.1. Thereby, the maximal convection velocity, which means the highest growth rate (in the case of precipitation, existent for the smallest nuclei), is the limiting factor.

2.1.2 Algorithms applied – Direct quadrature method of moments (DQMOM)

The basic idea of moment methods to solve the PBE, such as the quadrature method of moments (QMOM) [14] or direct DQMOM, is to solve the closure problem of moment methods by a quadrature approximation [15], where $\mathcal{M}_k$ is the kth moment of a PSD (eq. 2.4).

$$\mathcal{M}_k = \int_{-\infty}^{\infty} (L^P)^k f(L^P) dL^P \cong \sum_{i=1}^{N} w_i \mathcal{L}_i^{\,k} \qquad 2.4$$

Additional transported scalar equations must be solved within CFD. Thereby, the weights w_1, w_2 and the weighted abscissas $l_1 = w_1\mathcal{L}_1$ and $l_2 = w_2\mathcal{L}_2$ must be transported for the DQMOM approach with $N = 2$ nodes used in this paper. The source terms a_i and b_i are formulated according to the paper of Marchisio and Fox [16], including the moment source description for nucleation and growth (eq. 2.8 and 2.9) for the first four moments $k = 0, 1, 2, 3$ (detailed explanation in the appendix 8.5.1)

$$\frac{\partial \rho w_i}{\partial t} + \nabla \cdot (\vec{u} \rho w_i) = a_i(w_i, \mathcal{L}_i) \qquad 2.5$$

$$\frac{\partial \rho l_i}{\partial t} + \nabla \cdot (\vec{u} \rho l_i) = b_i(w_i, \mathcal{L}_i) \qquad 2.6$$

They can be received by a matrix inversion of a linear equation system of the form $\theta = \mathcal{B}^{-1}\psi$, where $\theta = [a_1, a_2, b_1, b_2]$ and $\psi = [S_0, S_1, S_2, S_3]$.

$$S_k = S_{\mathrm{Nuc},k} + S_{\mathrm{G},k} \qquad 2.7$$

$$S_{\mathrm{Nuc},k} = B_{\mathrm{hom}}(S, t) \cdot {L_{\mathrm{crit}}}^k \qquad 2.8$$

$$S_{\mathrm{G},k} = k \sum_{i=1}^{N} w_i \, {\mathcal{L}_i}^{k-1} G(\mathcal{L}_i) \qquad 2.9$$

For consistency with the multifluid model in CFD, the transport equations are associated with the appropriate scalars, such as α_i, the volume fraction of the solid phases and their effective length $\alpha_i \mathcal{L}_i$ [17] that must be transported by two additional user-defined scalars. The shape factor k_v is chosen to $\pi/6$ for a sphere. The procedure detailed is explained in the appendix or can be extracted from Liu and Fox [18].

$$\alpha_i = k_v w_i \mathcal{L}_i^3 = k_v \frac{l_i^3}{w_i^2} \qquad 2.10$$

$$\alpha_i \mathcal{L}_i = k_v w_i \mathcal{L}_i^4 = k_v \frac{l_i^4}{w_i^2} \qquad 2.11$$

Figure 2.3, left, shows the PSD calculated with the HRFVM (see 2.1.1) as a reference in comparison to the simulated finite node representation ($N = 2$ and $N = 3$) gained with the DQMOM algorithm.

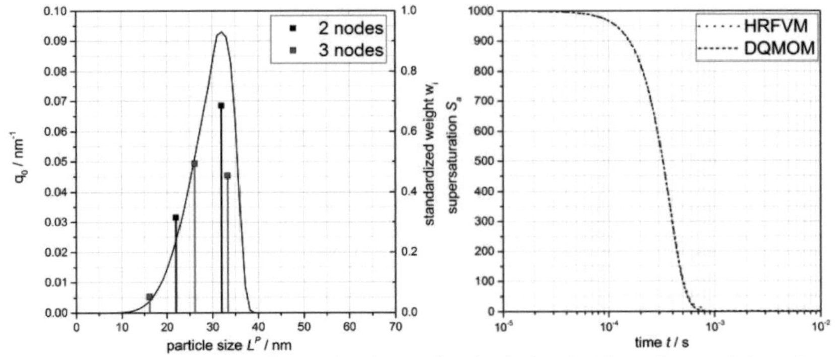

Figure 2.3: Left: Discrete particle size distribution (PSD) calculated with a reference finite volume method (HRFVM) and corresponding node representation ($N = 2$, black line) and ($N = 3$, dashed line), calculated with the DQMOM approach for ($S_{a,\mathrm{nom}} = 1000, \mathrm{R_c} = 5$, assuming perfect mixing). Right: Corresponding supersaturation depletion along the simulation time due to nucleation and growth for both algorithms tested.

Data are exemplarily calculated with kinetic approaches for nucleation and growth presented subsequently in section 2.3 for $BaSO_4$, assuming perfect mixing at t = 0. Moreover, when comparing the temporal progress of the depletion of supersaturation (Figure 2.3) simulated, one can see the mathematical congruency between the two very different population balance solver approaches.

The DQMOM solver represents a rather accurate possibility to model the PSD evolution within CFD with low computational costs compared to a discretely represented distribution composed of several size classes. It is fair to mention that the moment representation cannot be reconstructed easily into a PSD. Consequently, characteristic diameters should be chosen that can be calculated easily from the moments, such as a $L_{1,0}^P$ or a Sauter mean diameter $L_{3,2}^P$, for a comparison of experimental and simulative data.

2.2 Thermodynamic fundamentals

A thermodynamic mixed phase system with k components and a free surface is fully described by the Gibbs fundamental equation (eq. 2.12), which is the total differential of the internal energy U.

$$dU = TdS - pdV + \sum_k \mu_k dn_k + \sigma^{SL} dA \qquad 2.12$$

This equation may be quite inconvenient to work with since the two involved state variables S and V are not measurable directly. One can derive an expression of the Gibbs free energy which is much more applicable by Legendre transformation (eq. 2.13) [19].

$$dG = Vdp - SdT + \sum_k \mu_k dn_k + \sigma^{SL} dA \qquad 2.13$$

Therein, the involved intensive parameters p and T are quantities measurable directly. The chemical potential μ_i of a component i in a mixture phase is defined mathematically consonant with eq. 2.14 and describes the change of the Gibbs free energy, respectively, of the internal energy at a changing amount of component i at p and T= const.

$$\left(\frac{\partial G}{\partial n_i}\right)_{p,T,n_{j\neq i}} = \left(\frac{\partial U}{\partial n_i}\right)_{S,V,n_{j\neq i}} = \mu_i \qquad 2.14$$

The Gibbs free enthalpy of a multiphase system (e.g. phase α and β) can be calculated additively (eq. 2.15). Changes in the free enthalpy of the overall system have got to be zero in thermodynamic equilibrium (eq. 2.16).

$$G = G^\alpha + G^\beta \qquad 2.15$$

$$dG = dG^\alpha + dG^\beta = 0 \qquad 2.16$$

For p and T = const., considering eq. 2.13, the general constraint for phase equilibrium between a phase α and a phase β reads thus

$$\mu_i^\alpha = \mu_i^\beta \quad ; i = 1 \dots k \qquad 2.17$$

The chemical potential, an intensive state variable, as well as p and T of the liquid phase, are expressed as eq. 2.18, wherein a_i is the activity of component i in the mixture $(= \gamma_i \tilde{x}_i)$. Thereby, the activity coefficient $\gamma_i(T, \tilde{x}_i)$ is a temperature- and composition-dependent correction function that accounts for interactions of real systems. Pressure dependency plays only a negligible role and, therefore, is not commonly considered.

$$\mu_i = \mu_{0,i}(p,T) + RT\ ln\ a_i \qquad 2.18$$

The equation of state for the chemical potential is strongly dependent on the reference state chosen. Thereby, either the pure real component (index 0, see eq. 2.18) or the ideally dilute state (index ∞) of a solution is chosen for reference. If any component at p and T of the mixture is a gas or a solid, such as in a solvent solute equilibrium, the ideally dilute reference state has got to be chosen ($\tilde{x}_1 \to 1$: index 1 = solvent; $\tilde{x}_i \to 0$: i = 2...n). The chemical potential can then be expressed consonant with eq. 2.19, wherein $\gamma_i^r (= \gamma_i/\gamma_\infty)$ is named the rational activity coefficient. γ_∞ is the limiting activity coefficient at infinite dilution, which is $\neq 1$, independent of any concentration, but dependent on the composition of the mixture.

$$\mu_i = \mu_{\infty,i}(p,T) + RT\ ln\, \gamma_i^r \tilde{x}_i \qquad 2.19$$

The rational activity coefficient $\gamma_i^r \to 1$ for the limiting case of ideal dilution ($\tilde{x}_i \to 0$). In addition to the molar fraction $\tilde{x}_i$ in eq. 2.19, the chemical potential can be evaluated for different concentration quantities. In terms of molar concentrations $\tilde{c}_i$, the chemical potential can be written as eq. 2.20. Thereby, the appropriate reference potential formulation $\mu_{\infty,\tilde{c}_i}(p,T,\tilde{c}_{0,i})$ can be extracted from the appendix and from Schaber [19]. The reference concentration $\tilde{c}_i^0$ is typically chosen to be 1 mol/l. $\gamma_{\tilde{c}_i}^r$ in eq. 2.20 to get a dimensionless expression within the natural logarithm, and it is called the practical activity coefficient.

$$\mu_i = \mu_{\infty,\tilde{c}_i}(p,T,\tilde{c}_{0i}) + RT\ ln\, \gamma_{\tilde{c}_i}^r \frac{\tilde{c}_i}{\tilde{c}_{0,i}} \qquad 2.20$$

This practical activity coefficient can be converted into the rational one by $\gamma_{\tilde{c}_i}^r = \gamma_i^r \cdot \tilde{c}_\infty/\tilde{c}_{tot}$, whereby in an aqueous system ($\rho^L = 1000\ kg/m^3$) $\tilde{c}_\infty = \rho^L/\widetilde{M}_1^L = 55.56\ mol/l$.

The chemical potential for ionic systems is commonly expressed by use of molalities $\widetilde{m}_i$ (moles of component i/kg solvent) (eq. 2.21) and by use of a reference molality $\widetilde{m}_i^0 = 1\ mol/kg$.

$$\mu_i = \mu_{\infty,\widetilde{m}_i}(p,T,\widetilde{m}_{0,i}) + RT\,ln\,\gamma^r_{\widetilde{m}_i}\frac{\widetilde{m}_i}{\widetilde{m}_{0,i}} \tag{2.21}$$

The conversion between the two practical activity coefficients reads as eq. 2.22, wherein ρ^L is the solvent density.

$$\gamma^r_{\widetilde{m}_i} = \gamma^r_{\tilde{c}_i}\cdot\frac{\tilde{c}_i}{\widetilde{m}_i}\cdot\frac{1}{\rho^L} \tag{2.22}$$

2.2.1 Supersaturation

The fundamental driving force for crystallization and precipitation is the difference between the chemical potentials of a component i regarded in the transferring (e.g. solution) and transferred states (e.g. crystal) [20].

$$\Delta\mu_i = \mu_i^\alpha - \mu_i^\beta \tag{2.23}$$

In terms of precipitation, supersaturation is a transient and local quantity that is created either by the chemical reaction of components of the reactant solutions or through the reduction of the solubility by adding a drowning-out agent. The system regarded counteracts this thermodynamic non-equilibrium by segregation of a component i as a solid out of the liquid phase. The thermodynamic equilibrium is given by the solubility limit of the component i in the liquid phase. Thus, the effective driving force for crystallization is the difference between the chemical potential of the supersaturated component i and the one of the solid phase at the saturated state (see eq. 2.25). It is more appropriate for electrolyte solutions to use the mean ionic activity ($a = a_\pm{}^\nu$), since single electrolyte activity coefficients are experimentally inaccessible. The number of moles of ions in 1 mole of solute is ν $(= \nu_+ + \nu_-)$.

$$a_\pm{}^\nu = (a_{i+}^{v_+}\cdot a_{j-}^{v_-})^{1/(v_+ + v_-)} \tag{2.24}$$

The difference in the chemical potential of the system for component i can then be written as

$$\Delta\mu_i = \mu_i - \mu_i^* = \nu RTln(a_{\pm i}/a_{\pm i}^*) = \nu RTln(S_a) \tag{2.25}$$

The activity product of the equilibrium state is called the solubility product K_{sp}.

$$K_{sp} = a_{i+}^{*\,v_+} \cdot a_{j-}^{*\,v_-} = a_\pm^* \qquad 2.26$$

Combining eq. 2.24, 2.25 and 2.26, supersaturation can be expressed by

$$S_a = \exp\left(\frac{\Delta\mu_i}{RT}\right) = \left(\frac{a_{i+}^{v_+} \cdot a_{j-}^{v_-}}{K_{sp}}\right)^{1/(v_+ + v_-)}$$

$$= \gamma_{\tilde{c}_\pm}^r \left(\frac{\tilde{c}_{i+}^{v_+} \cdot \tilde{c}_{j-}^{v_-}}{K_{sp}}\right)^{1/(v_+ + v_-)} = \gamma_{\tilde{m}_\pm}^r \left(\frac{\tilde{m}_{i+}^{v_+} \cdot \tilde{m}_{j-}^{v_-}}{K_{sp}}\right)^{1/(v_+ + v_-)} \qquad 2.27$$

2.2.2 Free lattice ion ratio

Another important parameter that will be used in following sections is the free lattice ion ratio R_c that accounts for the stoichiometric ratio of reacting ions (see eq. 2.28). Considering eq. 2.27, one can easily recognize that identical supersaturation values can be adjusted by different combinations of the ions involved. Therefore, this second parameter is necessary to provide for distinct precipitation conditions.

$$R_c = \frac{\tilde{c}_{i+}}{\tilde{c}_{j-}} = \frac{\tilde{c}_{Ba^{2+},\text{free}\cdot}}{\tilde{c}_{SO_4^{2-},\text{free}}} \qquad 2.28$$

The factor R_c for the model system $BaSO_4$ is formulated according to Kügler et al. [3], Kucher et al. [21] and Schwarzer [5], as shown in eq. 2.28, considering the concentration of freely solvated ions that are not bound by ion complexes (see also section 2.2.1). This expression is one of several possibilites to express the ratio of reacting components involved. Hence, in terms of comparability to literature, R_c will be used in this work. But, and this should be scrutinized, either an activity-based formulation R_a or even a hypothetic, crystal-surface specific adsorptive ratio would be a more appropriate quantity. However, these dependencies are not investigated in this work.

2.2.3 Activity coefficient approaches

Activity coefficients are the correction factors that account for deviations from real to ideal behavior of a mixture. These deviations in ionic aqueous mixtures, as regarded within this work, can arise from ion-interactions and ion-associations [22, 23].

Activity coefficients play an important role for the proper description of the driving force for primary processes at precipitation, such as nucleation, which is strongly nonlinear (see section 2.3.1). A simple exemplary comparison can highlight this matter well and embodies the importance of the activity coefficient approaches for process simulations (see Table 2.1).

Table 2.1: Comparison of supersaturation values calculated by different activity coefficient approaches for $BaSO_4$ with $R_c = 1$.

S_c	$S_{a,Truesdell-Jones}$	$S_{a,Bromley}$	$S_{a,Pitzer}$
5245	716	875	800
8975	989	1158	1000

Certain models with varying complexity have been proposed in literature for ionic systems to predict such coefficients [22, 23, 24]. Model accuracy is often only satisfactory under dilute conditions, and decreases with the excess of one reactant which may occur during the mixing step [7]. The approaches usually available are based upon calculation of the molal practical activity coefficient $\gamma^r_{\tilde{m}_i}$ (see eq. 2.21). In terms of crystallization applications, molality offers certain advantages, since the reference quantity of the solvent mass contrasts to a specific volume (e.g. $[\tilde{c}_i] = \text{mol/m}^3$) not dependent on temperature or pressure.

The most influential quantity in the activity models listed is the ionic strength I of the solution. Thereby, the approaches are set up based on the molal ionic strength $I_{\tilde{m}}$ (eq. 2.29). The use of a concentration based ionic strength is permitted for the dilute case and delivers similar results.

$$I_{\tilde{m}} = 0.5 \sum_k \tilde{m}_k z_k^2 \qquad 2.29$$

Table 2.2 presents an overview of several activity models provided in literature. They are usually built on the Debye-Hückel approach, which is itself only valid for highly dilute solutions [23, 6]. Either strongly simplified thermodynamic models or empirical expansions that involve ion specific interaction parameters which must be taken from tabulated literature are used. It should be considered that every approach has its own limit of validity.

Table 2.2: Overview of several activity coefficient models from literature

Model	Basic equation	Validity $I_{\tilde{m}}$
Debye-Hückel [22]	$\log \gamma^r_{\tilde{m}_\pm} = -A_{\mathrm{DH}} \lvert z_{i+} z_{j-} \rvert \sqrt{I}$	$< 10^{-2}\ M$
Extended Debye-Hückel [6]	$\log \gamma^r_{\tilde{m}_\pm} = -A_{\mathrm{DH}} \lvert z_{i+} z_{j-} \rvert \left(\frac{\sqrt{I}}{1 + B_2 a_{eff,i} \sqrt{I}} \right)$	$< 0.1\ M$
Davies [25]	$\log \gamma^r_{\tilde{m}_\pm} = -A_{\mathrm{DH}} \lvert z_{i+} z_{j-} \rvert \left(\frac{\sqrt{I}}{1 + \sqrt{I}} - 0.3 I \right)$	$< 0.5\ M$
Truesdell-Jones (WATEQ) [26]	$\log \gamma^r_{\tilde{m}_\pm} = -A_{\mathrm{DH}} \lvert z_{i+} z_{j-} \rvert \left(\frac{\sqrt{I}}{1 + B_2 a_{eff,i} \sqrt{I}} \right) + b_i I$	$< 1\ M$
Bromley [22]	$\log \gamma^r_{\tilde{m}_\pm} = -A_{\mathrm{DH}} \lvert z_{i+} z_{j-} \rvert \left(\frac{\sqrt{I_{\tilde{m}}}}{1 + B_2 a_{eff,i} \sqrt{I_{\tilde{m}}}} \right) + \frac{1}{2} (F_{i+} + F_{j-})$	$< 6\ M$
Pitzer [23]	$\ln \gamma^r_{\tilde{m}_i} = \frac{\partial \left(\frac{G^E}{RT} \right)}{\partial N_i}$ $\frac{G^E}{m_{\mathrm{sol}} RT} = f(I_{\tilde{m}}) + \sum_i \sum_j \lambda_{ij}(I_{\tilde{m}}) \tilde{m}_i \tilde{m}_j + \sum_i \sum_j \sum_k \mu_{ijk}(I_{\tilde{m}}) \tilde{m}_i \tilde{m}_j \tilde{m}_k$	$> 6\ M$

The focus in this work is laid on the Pitzer approach, which is, due to the consideration of binary as well as ternary interaction parameters, supposed to be the most accurate model available. Since no Pitzer parameters are available for some substances investigated, such as SrF_2 (subsequently, see chapter 3), the Truesdell-Jones model acts a makeshift. A detailed, well explained formulation of the Pitzer approach for $BaSO_4$ can be found in Steyer et al. [24].

Moreover, in terms of the focused model system barium sulfate, it is considered that the system does not behave like a strong electrolyte which is fully dissociated in water. It has been shown by Felmy et al. [27], Monnin [28] and Vicum et al. [29] that $BaSO_4$ forms undissociated ion-pair complexes. The equilibrium corresponding is described consonant with eq. 2.30.

$$K_{IP} = \frac{\widetilde{m}_{i+}\widetilde{m}_{j-}\gamma^{r}_{\widetilde{m}_{\pm}}{}^{2}}{\widetilde{m}_{ij(aq)}\,\gamma^{r}_{\widetilde{m}_{ij}(aq)}} \qquad 2.30$$

The equilibrium constant is taken from Felmy et al. [27] ($K_{IP} = 10^{-2.26}$). The objective of this thesis later on (see section 3) is to elaborate a flow sheet tool that is able to handle the hydrochemistry of reacting metal salts with a high degree of flexibility. Since the solubility equilibrium of complex ionic systems is often influenced by a multitude of complexes and different phase equilibria which affect the real behavior of the solution, the open source hydrochemistry software *PhreeqC* (*U.S. Geological Survey*) is used as an aid. The software (based on C++) provides certain model systems, such as the Pitzer approach or the Truesdell-Jones model.

Figure 2.4 shows a typical precipitation diagram for the reactants Na_2SO_4 and $BaCl_2$ wherein three approaches (namely Bromley, Pitzer and Truesdell-Jones) are calculated. The Bromley and Pitzer model are calculated manually consonant with Steyer et al. [24].

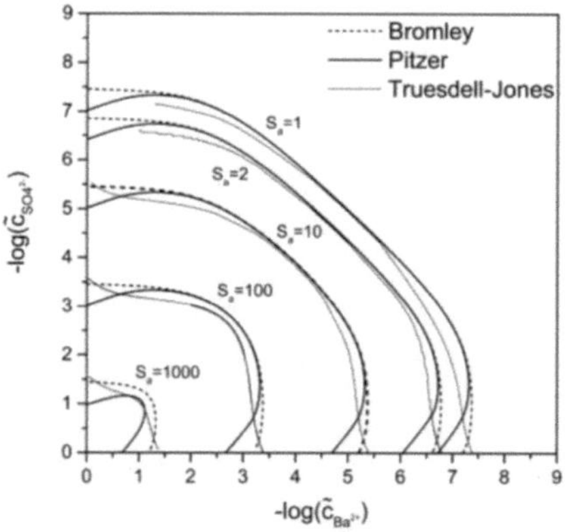

Figure 2.4: Precipitation diagram, involving three different activity coefficient approaches. The lines mark different combinations of ions leading to the equal supersaturation. The bisecting line in the diagram depicts the stoichiometric case ($R_c = 1$*).*

The Truesdell-Jones lines are calculated using the PhreeqC software. As one can see for the stoichiometric case ($R_C = 1$, bisecting line), the three approaches deliver quite similar results which differ more strongly for higher supersaturations. All three models show different trends for the highly nonstoichiometric case. Since the Pitzer approach involves most interaction parameters and is defined for the broadest range of validity, this model is used in the following for barium sulfate.

The activity models which are used need the conversion from molar concentrations, which are balanced in combination with the population balance into molalities, to calculate an activity coefficient. Literature [4, 5] often relinquishes this conversion, since the exact density of the mixture is mostly unknown. PhreeqC executes the conversion internally, consonant with eq. 2.31, with a fixed mixture density ρ^L of $1000\ kg/m^3$.

$$\widetilde{m}_i = \frac{\tilde{c}_i}{\rho^L - \sum_k \tilde{c}_k \widetilde{M}_k} \tag{2.31}$$

Figure 2.5 shows the differences arising when molar concentrations are used instead of molalities. On the one hand, the parity plot shows the deviations for a fixed density used within *PhreeqC*.

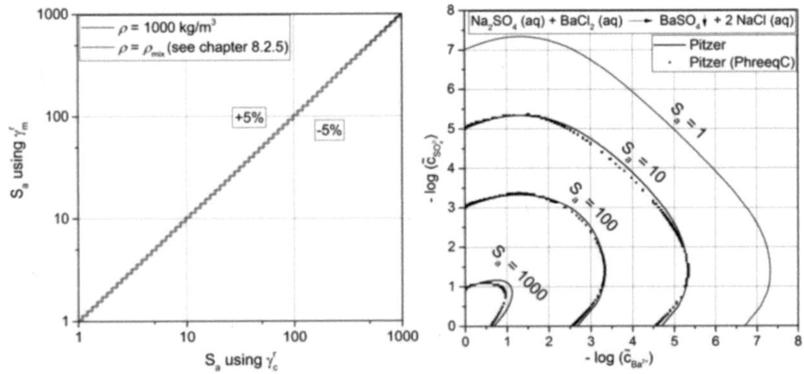

Figure 2.5: Left: Parity plot investigating the error potential using a concentration-based (Pitzer approach) instead of a molality-based activity coefficient. Right: Comparison of the Pitzer model predictions calculated manually consonant with Steyer et al. [24] and calculated with PhreeqC.

On the other hand, the mixture density model for the two aqueous ionic reactant solutions Na_2SO_4 and $BaCl_2$ (see appendix 8.2.5 or Metzger and Kind [30]), which is developed subsequently especially for the CFD application, is inserted for ρ^L in eq. 2.31.

Thereby, the deviations can be seen to be tolerable ($\ll 5\ \%$) for a high quantity of ions dissolved (high supersaturations). Figure 2.5, right, shows the comparison between results obtained with the Pitzer approach calculated manually and those from the *PhreeqC* software. *PhreeqC* uses a slightly different solubility product ($K_{sp} = 1.07 \cdot 10^{-10}\ mol^2/l^2$ instead of $9.82 \cdot 10^{-11}\ mol^2/l^2$ [28]) and does not involve the ion pair complex $BaSO_{4(aq)}$ formation, addressed in the section before. However, comparatively congruent results are obtained, which confirm that the open source hydrochemistry solver is a good substitute for a flexible tool.

2.3 Precipitation

2.3.1 Nucleation

Nucleation is the initial event which must occur to create a crystalline solid out of the liquid phase. Thereby, supersaturation, respectively, a gradient in chemical potentials (see eq. 2.25), serves as the driving force for this primary process of particle synthesis. Nucleation may occur spontaneously or be induced artificially, for example, via shear, attrition or contact [20].

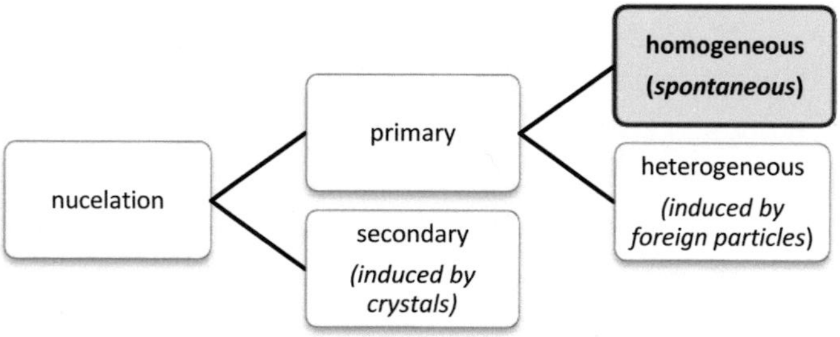

Figure 2.6: Basic terminological classification of nucleation, taken from Mullin [20], whereby the secondary nucleation can be induced by a group of mechanisms, such as shear, attrition, fracture or contact [8], which will not be regarded in detail here.

Figure 2.6 introduces the basic terminological classification of nucleation which is subdivided into a primary and a secondary type.

This work focuses solely on homogeneous nucleation as a subbranch of primary nucleation which occurs spontaneously out of the supersaturated solution. This phenomenological limitation of mechanisms should be appropriate due to high supersaturation values generated by the chemical reaction in the precipitation process [8, 20]. Suitable literature to gain a diversified overview and to deepen insight into specific nucleation mechanisms is given by Mersmann [8] and Kashchiev [31].

The classical nucleation theory (CNT) was introduced by Becker, Döring, Volmer, Gibbs and others and is based originally on the condensation of a vapor to a liquid and, subsequently, extended to crystallization from solutions [1, 8]. Thereby, homogeneous nucleation in a metastable, supersaturated liquid phase is introduced by the formation of molecular clusters (with surface A^P and volume V^P). The overall excess free energy ΔG, between such a small cluster and the solute in solution is obtained by balancing the free-surface energy needed to build this new surface ΔG_A and the free energy ΔG_V (proportional to the volume of the cluster) that is gained by the formation the solid phase. The summarized potential function reads as eq. 2.32, wherein the molar volume is defined as $V_m = \widetilde{M}^S/(\rho^S N_a)$.

$$\Delta G = \Delta G_A + \Delta G_V = A^P \gamma^{SL} - \frac{V^P}{V_m} k_B T \nu ln S_a \qquad 2.32$$

Depending on L^P, ΔG passes through a maximum at a distinct (critical) cluster size L^P_{crit} [32, 20]. Clusters that reach the size of this threshold value are considered to act as nuclei which continue to grow, whereas smaller clusters below L^P_{crit} disintegrate again [5].

The critical nuclei size at a given supersaturation S_a and temperature T can then be calculated by the search for the extreme value of eq. 2.32 $(\partial \Delta G/\partial L = 0)$ and reads as eq. 2.33 (see also Figure 2.7).

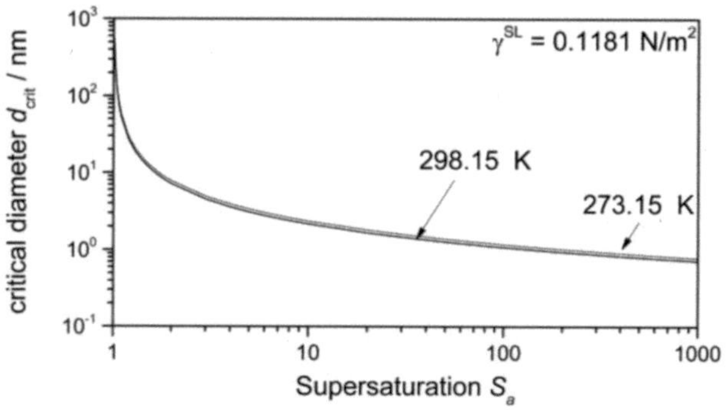

Figure 2.7: Critical nuclei size for different supersaturation values according to eq. 2.33 for $\gamma^{SL} = 0.1181\ N/m^2$, which is the interfacial energy of $BaSO_4$-water, taken from [4].

$$L_{\text{crit}}^{P} = \frac{4\gamma^{SL}V_m}{\nu k_B T ln S_a} \qquad 2.33$$

The rate of nucleation within the CNT is expressed as the rate of clusters that cross the critical energy barrier ΔG_{crit} mentioned above. Thereby, it can be calculated as a product of three factors that contribute (see eq. 2.34).

$$B_{\text{hom}} = \beta_c \cdot n_{\text{crit}} \cdot Z \qquad 2.34,$$

where n_{crit} is the number concentration of critical clusters caused by random collisions of molecules, which can be described by a Boltzmann distribution (see eq. 2.35) [8]

$$n_{\text{crit}} = n_0 \cdot e^{(-\Delta G_{\text{crit}}/k_B T)}, \qquad 2.35,$$

where n_0 is the number concentration of monomers in the supersaturated solution. Consonant with Kucher and Kind [4] and Schwarzer [5], the following expression is chosen.

$$n_0 = (a_{i+}^{v_+} \cdot a_{j-}^{v_-})^{\frac{1}{v_+ + v_-}} \cdot N_a = \sqrt{K_{sp}} \cdot S_a \cdot N_a \qquad 2.36$$

The free nucleation enthalpy of the critical cluster can be calculated according to eq. 2.37, inserting L_{crit}^{P} into the expressions of ΔG_A and ΔG_V in eq. 2.32. For more details, see Mersmann [8].

$$\Delta G_{\mathrm{crit}} = \frac{16\,\pi (\gamma^{SL})^3 {V_m}^2}{3\,(k_{\mathrm{B}} T \ln S_a)^2} \tag{2.37}$$

The second factor in eq. 2.34, β_c, which is called the collision rate. Several formulations exist for this factor in literature. This work follows the approach of Kind and Mersmann [33], who derive their expression considering the nucleation of condensed fluids.

$$\beta_c = 3/4 \cdot n_0^{4/3} \cdot D_{AB}^L \cdot A_{\mathrm{crit}} \tag{2.38}$$

The third factor involved in eq. 2.34 represents the Zeldovich factor which is an imbalance factor calculated from the second derivative of the free enthalpy at ΔG_{crit}. It accounts for the non-equilibrium cluster size distribution due to the steady removal of those clusters that cross the potential barrier ($L^P \geq L_{\mathrm{crit}}^P$) [31]. For more information on the derivation of eq. 2.39, Kashchiev [31] and Mersmann [8] are recommended.

$$Z = \frac{2 \cdot V_m}{A_{\mathrm{crit}}} \sqrt{\frac{\gamma^{SL}}{k_{\mathrm{B}} T}} \tag{2.39}$$

Combining eq. 2.34 – 2.39, one gets eq. 2.40, which is the expression of the CNT used in this work.

$$B_{\mathrm{hom}} = \frac{3}{2} D_{AB} \cdot \left(\sqrt{K_{sp}} \cdot S_a \cdot N_a\right)^{\frac{7}{3}} \cdot \sqrt{\frac{\gamma_{SL}}{k_{\mathrm{B}} T}} \cdot V_m \, exp\left(-\frac{16\pi}{3} \cdot \left(\frac{\gamma^{SL}}{k_{\mathrm{B}} T}\right)^3 \cdot \frac{V_m^2}{(\nu ln S_a)^2}\right) \tag{2.40}$$

One uncertainty in eq. 2.40 is the interfacial tension between the crystalline nucleus and its surrounding liquid. Since the interfacial energy is not a measurable quantity, it must be taken as a fit parameter from primary nucleation rate measurements. Consonant with Mersmann [8], the interfacial energy can be described reasonably well by eq. 2.41.

$$\gamma^{SL} = K_0 \cdot k_B T \cdot {V_m}^{-2/3} \cdot \ln \frac{\rho^S}{\widetilde{M}^S \cdot \sqrt{K_{sp}}} \tag{2.41}$$

A value of 0.414 is recommended for the empirical factor K_0, which is a fit constant gained from the adaption of γ^{SL} to experimental data taken from 58 systems investigated by Nielsen and Söhnel [34] and Söhnel [35, 36]. Mersmann [8] mentions that the accuracy of eq. 2.41 may be improved by considering some occurring value-diminishing surface charge effects. Such changes in the surface charge density, which influence γ^{SL} directly, are caused by either the adsorption of potential-determining lattice ions or an increased ionic strength $I_{\tilde{m}}$ of the mother liquor, influencing the thickness of the electrical double layer. Schwarzer [5], for instance, considered these contributions for the system $BaSO_4$–water and calculated for a nominal supersaturation of $S_a = 1000$ and an excess of barium-ions $\mathrm{R_c} = 5$ instead of a value of $0.1284\ N/m^2$ (eq. 2.41) a value of $\gamma^{SL} = 0.1181\ N/m^2$.

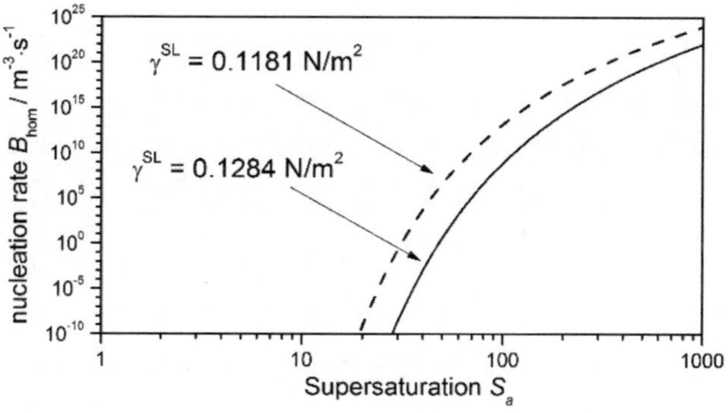

Figure 2.8: Strongly nonlinear behavior of the homogeneous nucleation rate independent of the supersaturation S_a and γ^{SL} for the model system $BaSO_4$ [4, 5] ($K_{sp} = 9.81 \cdot 10^{-11} mol^2/l^2$) [28].

2.3.2 Growth

Crystal growth of ionized solutes can be considered as a highly complex process that is not well understood in specific ways [8]. A number of subsequent steps have to take place until a hydrated growth unit (e.g. either an ion or a larger cluster of ions) is incorporated within the crystal lattice. Mullin [20] and Mersmann [8] give an overview of consecutive and parallel steps that have to take place. Thus, growth involves processes such as ion diffusion from the bulk through the diffusion and adsorption layers to the crystal surface, surface diffusion of ions (solvated and unsolvated) and their desolvation before becoming integrated into the crystal surface. These mechanisms are accompanied by the counter-diffusion phenomena of water as well as surface charge- and counter-ion diffusion effects in the nonstoichiometric case. The slowest of these steps should be rate determining. However, growth can be subdivided basically into the transport and the integration step of ions.

Growth is controlled by diffusion from the bulk towards the crystal for high supersaturation regarding the precipitation processes within this work [8]. Hence, it is assumed that ions that are carried to the crystal surface are incorporated directly into the crystal structure. The experimental findings of Li et al. [37] and Soleymani et al. [6] support such considerations. A change in the growth mechanism is observed firstly for relatively low saturation values ($S_a = 40$). This section focuses only on this diffusion-limited case. A detailed description of integration-controlled growth mechanisms can be found in Mersmann [8] and Mullin [20].

The work of Schwarzer [5] and Kucher and Kind [4], who deduce the governing source term formulation of particle growth from a linear mass transport approach (Fick's law), is followed in this section.

$$\dot{m}_i = \beta_i \cdot A \cdot \widetilde{M}_i \cdot (a_i - a_i^*) \qquad 2.42$$

The mass transfer coefficient can be substituted using the definition for the Sherwood number $Sh\,(= \beta \cdot L^P / D_{AB}^L)$. Mersmann [8] gives a Sherwood correlation for a suspended particle in a turbulent flow (eq. 2.43, Figure 2.9) independent of the mean energy dissipation $\bar{\varepsilon}$. Since no flow field-induced relative motion exists for small sphere-like particle, $\mathrm{Sh} = \mathrm{Sh}_{\mathrm{min}} = 2$.

$$\mathrm{Sh} = \mathrm{Sh}_{\mathrm{min}} + 0.8 \cdot \left(\frac{\bar{\varepsilon} \cdot (L^P)^4}{\nu_{\mathrm{kin}}^3}\right)^{1/5} \cdot \left(\frac{\nu_{\mathrm{kin}}}{D_{AB}^L}\right)^{1/3} \qquad 2.43$$

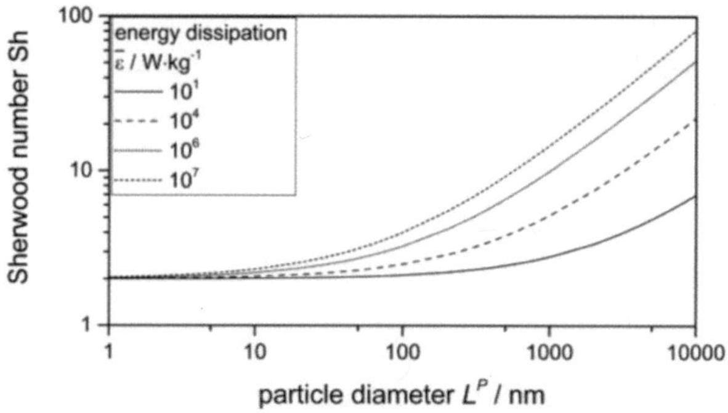

Figure 2.9: Sherwood number for a sphere-like crystal ($\mathrm{Sh_{min}} = 2$), calculated according to eq. 2.43 for several mean energy dissipation rates. $D_{AB}^{L} = 10^{-9}\, m^2/s; \nu_{\mathrm{kin}} = 1.0018 \cdot 10^{-6}\, m/s$).

The total mass flux kinetic can be expressed as eq. 2.44.

$$\dot{m}_{\mathrm{tot}} = \sum_i \dot{m}_i = \beta \cdot A \cdot \sum_i \left(\widetilde{M}_i \cdot (a_i - a_i^*)\right) \tag{2.44}$$

The mass flux towards a particle with the uniform density ρ^P and a volumetric shape factor k_V can be expressed with eq. 2.45, which leads to a term for the diffusion-limited growth rate G_{diff} (eq. 2.46)

$$\begin{aligned} \dot{m}_{\mathrm{tot}} = \frac{dm^P}{dt} = \rho^P \cdot \frac{dV^P}{dt} = \rho^P \cdot \frac{dV^P}{dL^P} \cdot \frac{dL^P}{dt} = \rho^P \cdot k_V \frac{d((L^P)^3)}{dL^P} \cdot \frac{dL^P}{dt} \\ = 3 \cdot k_v \cdot \rho^P \cdot (L^P)^2 \cdot G_{\mathrm{diff}} \end{aligned} \tag{2.45}$$

$$G_{\mathrm{diff}} = \frac{\dot{m}_{\mathrm{tot}}}{3 \cdot k_v \cdot \rho^P \cdot (L^P)^2} \tag{2.46}$$

Combining eq. 2.44 with eq. 2.46 ($A^P = k_A \cdot (L^P)^2$) leads to eq. 2.47.

$$G_{\mathrm{diff}} = \frac{k_A}{k_v} \cdot \frac{Sh \cdot D_{AB}^L}{3 \cdot \rho^P \cdot L^P} \sum_i \left(\widetilde{M}_i \cdot (a_i - a_i^*)\right) \tag{2.47}$$

Assuming sphere-like crystals with the volumetric and surface shape factors $k_v = \pi/6$ and $k_a = \pi$, one gets eq. 2.48.

$$G_{\mathrm{diff}} = 2 \cdot \frac{Sh \cdot D_{AB}^L}{\rho^P \cdot L^P} \sum_i \left(\widetilde{M}_i \cdot (a_i - a_i^*)\right) \tag{2.48}$$

The diffusion coefficient D_{AB}^{L} is calculated using the Stokes-Einstein equation, which is supposed to be a fairly good approximation. Comparison of tabulated ion diffusion coefficients and the ones calculated with eq. 2.49 can be found in the appendix.

$$D_{AB}^{L} = \frac{k_B \cdot T}{3\pi \cdot \eta^L \cdot r_{h,AB}} \qquad 2.49$$

Literature [5, 4] often provides a further simplification assuming equal activity differences as, for example, in stoichiometric precipitation without asymmetric electrolyte behavior (eq. 2.51). $\tilde{V}_m$ is the volume of one mole of the salt AB $(= \tilde{M}_{AB}/\rho^P)$.

$$\sum_i \left(\tilde{M}_i \cdot (a_i - a_i^*)\right) = \tilde{M}_{AB} \cdot \sqrt{K_{sp}} \cdot (S_a - 1) \qquad 2.50$$

$$G = Sh \cdot \frac{2D_{AB}\tilde{V}_m}{L^P} \cdot \sqrt{K_{sp}} \cdot (S_a - 1) \qquad 2.51$$

Schwarzer [5] discusses this simplification critically and proposes – in addition to the stoichiometric case – using eq. 2.51 as the general form. The justification made is plausible.

Assuming, for instance, $BaSO_4$ precipitation with a barium excess (which is chosen for most of the experiments in this work), the activity gradient from the bulk to the crystal surface is larger for Ba^{2+} ions than the one for the SO_4^{2-} ions. It is assumed that a higher flux of Ba^{2+} leads subsequently initially to a buildup of surface charge which decelerates the Ba^{2+} ion rate and enhances the transport of SO_4^{2-}. This barium-ion-repulsive charge is built up much faster than the overall growth process lasts. Thus, a pseudo steady-state condition and an equilibrium surface charge for the mass transport is reached which is additionally superposed by charge effects (attraction and repulsion). Considering equal transport rates for the ions, which is needed to build the crystal lattice, the mean transport rate of ions is smaller than the transport rate for Ba^{2+} and larger than the one for the SO_4^{2-}-ions in deficit. Schwarzer [5] could show, by considering these charge effects within an E-field approach, that eq. 2.51 agrees fairly well with the results of the more complex solution.

2.4 Simulation assumptions using $BaSO_4$ as a model system

This section introduces some state-of the-art knowledge which is necessary to assess data achieved in this work. Moreover, essential assumptions and restrictions that are made for the simulation of precipitation processes especially for the model system $BaSO_4$ are presented in the following. Findings are supposed to be reasonably well transferable to other inorganic systems. The extent of validity of this combination of approach will be discussed in section 3. The basic simulation assumptions made read as follows.

1) Supersaturation must be calculated activity based involving nonstoichiometric dependencies.

Supersaturation, as the driving force for nucleation and growth (see section 2.3.1 and 2.3.2), plays a fundamental role. When choosing, for example, a concentration-based expression for the supersaturation, nucleation is significantly overrated (see Table 2.1). Moreover, for strongly nonstoichiometric cases, this imbalance must be carefully considered in the activity coefficient models. Details are exposed in section 2.2 and will also be discussed later in section 3.

2) A stoichiometric ratio of $\mathrm{R_c} \geq 5$ is considered for the experiments and for the simulations to prevent agglomeration.

Electrostatic stabilization can be achieved for specific reactant ratios, expressed with a free, concentration-based, lattice ion ratio $\mathrm{R_c}$ (eq. 2.28), as shown in Kucher and Kind [4] and Schwarzer [5].

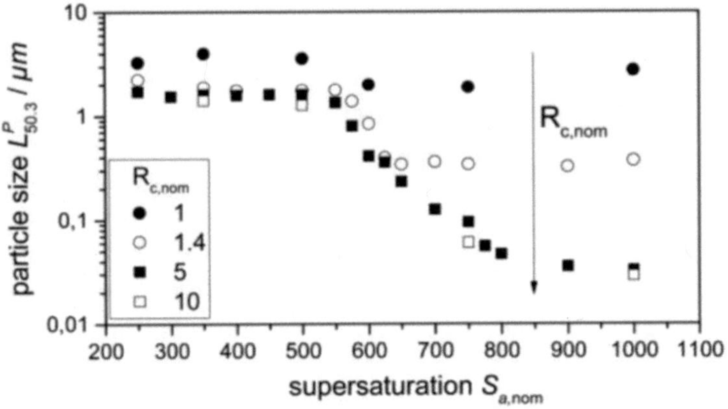

Figure 2.10: Influence of the nominal supersaturation $S_{a,\mathrm{nom}}$ and the nominal free ion ratio $\mathrm{R}_{\mathrm{c,nom}}$ on the particle size measured. Data are extracted from Kucher and Kind [4].

The latter authors demonstrate that behavior using an excess of the potential-determining ion-type barium ($\mathrm{R}_{\mathrm{c,nom}} \geq 5$) electrostatically stabilized primary particles can be formulated (see Figure 2.10).

3) Classical nucleation theory and diffusion-limited growth are adequate approaches to model the primary processes of particle formation.

A decision was made to follow the CNT consonant with Vollmer and Weber [8], which originally describes the condensation of vapor to a liquid droplet. Thereby, it is assumed that this theory can be transferred to crystallization from liquid, as it is done in manifold ways in literature [8]. It should be mentioned that literature provides different approaches, such as non-classical nucleation theory, where a metastable transition phase exists, or some empirical models, which are inherently dependent on the system investigated. We assume that also in the case of a metastable transition zone, which is observed for some material systems [38], the CNT does not lose its validity. See section 2.3.1 for details.

4) Diffusion-limited aggregation and dendritic growth are not considered.

Barium sulfate, shows a characteristic non-steady trend (see bend in Figure 2.10 and Figure 2.11 at $S_{a,\mathrm{nom}} \approx 550$) in the mean particle size $L^P_{50,3}$ and $L^P_{50,0}$ measured, even for $\mathrm{R_c} \geq 5$. This bend is nonexistent in simulation predictions using the theoretical particle formation approaches – CNT and diffusion-limited growth (see Figure 2.11, black line). Kügler and Kind [3] could show by adding a dispersing surfactant (MelPers0045, which is a polycarboxylate ether-based superplasticizer (PCE)) to the reactant solution, that final mean particle sizes can be measured following strictly the theoretically postulated ones using the kinetic approaches presented in section 2.3.1 and 2.3.2 (see Figure 2.11). It is not fully clarified whether the surfactant used suppresses the aggregation of the primary particles or the dendritic growth which may occur for lower supersaturation values ($S_{a,\mathrm{nom}} < 800$) [6], or both of them. However, the particle sizes measured and scanning electron microscope (SEM) images made by Kügler et al. [3] are important evidence that support the hypothesis that pure primary processes of particle formation, when aggregation and dendritic growth are suppressed, follow the kinetics presented.

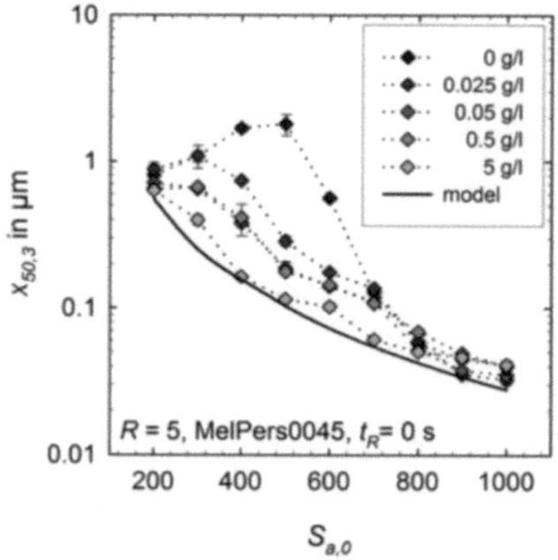

Figure 2.11: Influence of a surfactant (MelPers0045) on the particle size measured. The simulated line "model" uses the approaches presented in chapter 2.3 (CNT, diffusion-limited growth, Bromley activity coefficient model) for the case of ideal mixing at $t = 0$, cited from Kügler et al. [3].

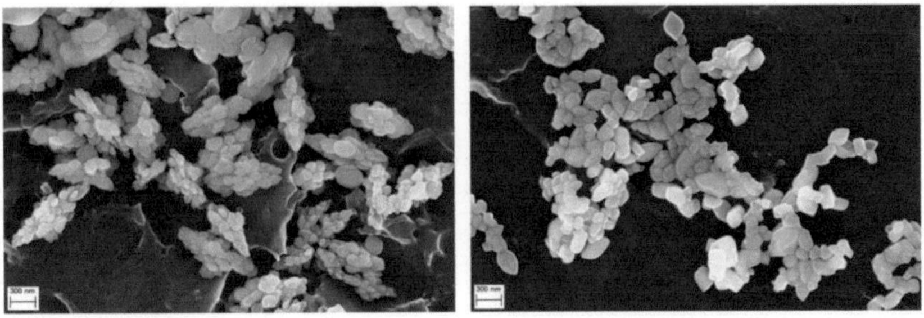

Figure 2.12: Scanning electron microscope (SEM) images of $BaSO_4$ precipitated in a Y-CIJM without mixing influence at $S_{a,nom} = 600$ and $\mathrm{R_c} = 5$, left: Without and right: With surfactant, cited from Kügler et al. [3].

Further important evidence regarding the validity of the approaches applied will be given in section 3.4, where complex, temporally superimposed associations lead to correct particle size predictions.

Since these fundamentals of particle formation still deserve further investigation, which is not part of this work, aggregation phenomena and dendritic growth are not considered. The influence of mixing is investigated for $S_{a,\mathrm{nom}} = 1000$ and $R_{c,\mathrm{nom}} = 5$, where such effects are not observed Thus, one is able to compare simulated and experimental data to each other.

5) Fick's law is assumed to be a fairly good approximation for the unidirectional diffusion-limited mass transfer during growth.

Diffusion-limited growth of a particle is a highly complex topic, as explained in section 2.3.2. From a general point of view, unidirectional mass transfer should be described with a Stefan-Maxwell mass transfer approach. Mass transport needs to consider additional contributions for concentrated electrolyte systems. The corresponding general Stefan-Maxwell expression in a molar concentration-based formulation reads according to Taylor and Krishna [39] as eq. 2.52 (u^L =solvent velocity, $i = Na^+, SO_4^{2-}, Ba^{2+}, Cl^-, H_2O$), where $\dot{n}_i$ represents the flux of each species i ("Extended Nernst-Planck model"). $[\mathrm{D_{ij}^L}]$ represents the diffusion coefficient tensor of component i in the solvent j. For further details the contribution of Samson and Marchand [40] is recommended.

$$\vec{\dot{n}_i} = -\left[D_{ij}^L\right]\left(\nabla\tilde{c}_i + \frac{z_i\mathcal{F}}{RT}\tilde{c}_i\nabla\phi_e + \tilde{c}_i\nabla(ln\,\gamma_i)\right) + \tilde{c}_i\overrightarrow{u^L} \tag{2.52}$$

$$\sum_{i=1}^{n} z_i\,\nabla\tilde{c}_i = 0 \tag{2.53}$$

The three contributions in eq. 2.52 are diffusion, migration due to the electrical potentials of the ions and convection. Equation 2.53 represents the electroneutrality condition that acts as a closure. This set of equations is highly complex to solve. Samson and Marchand [40] give an idea on how to progress such a mathematically challenging problem. Moreover, the crystal surface charge during nonstoichiometric precipitation should be considered additionally [5]. Since Kucher et al. [21] and Schwarzer [5] were able to predict particle sizes highly congruent to experiments by use of the significantly simpler linear mass transfer approach (eq. 2.51), this work also supposes that Fick's law should be an adequate choice. Temporal evidence found subsequently in section 3.4.3 strengthens the validity of this assumption.

6) Agglomeration and aging can be neglected for the electrostatically stabilized case ($\mathrm{R_c} \geq 5$).

As discussed in point 2), all experiments and simulations are made at $\mathrm{R_c} \geq 5$ (barium excess).

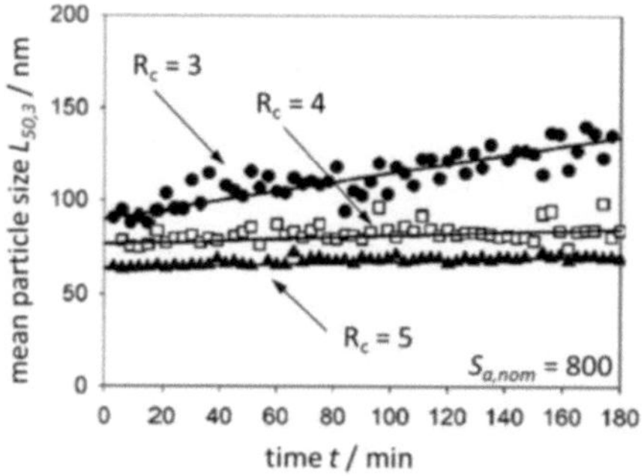

Figure 2.13: Measured temporal evolution of mean volumetric particle sizes in $BaSO_4$ suspensions after precipitation at $S_{a,nom} = 800$ (Bromley activity coefficient model used) at varying free lattice ion ratio $\mathrm{R_c}$ (figure modified from Kucher and Kind [4])

Kucher and Kind [4] could show, at least for the time range observed ($180\ min$), that no agglomeration due to attracting forces (e.g. van der Waals) should be considered for this electrostatically stabilized case. Experimental results are shown in Figure 2.13.

7) Attrition and breakage phenomena can be neglected for the particle sizes regarded.

The size of the particles formulated range between $\sim 30\ nm$ to $3\ \mu m$. Mersmann [8] shows convincingly that attrition and breakage only play a role for particles above $300\ \mu m$ in ST reactors even at a high specific power input. This threshold value is a conservative one, since the effect has been demonstrated for KNO_3 crystals that tend to be prone to such phenomena.

The region of high energy dissipation ($\bar{\varepsilon}_{\mathrm{CIJM}} \gg \bar{\varepsilon}_{\mathrm{ST}}$) in the CIJMs investigated only involves newly built particles in the range of a few nanometers that are assumed to follow the flow without extra drag (St $\ll 1$) (see also chapter 3.4.2).

8) The variation of the ionic strength and temperature which influence aggregation phenomena is not investigated within this thesis

Simulations are executed only under standard conditions ($\vartheta = 25\ °C, 1\ bar$). The effect of varying the ionic strength by a surplus of inertial ions influences mainly the Debye length and, thus, the agglomeration tendency of the particle collective. Increasing the temperature also supports this tendency, since the mobility of ions and, thus, their collision frequency is raised. Such effects are not regarded, because aggregation and agglomeration are undesirable phenomena when investigating the influence of mixing. However, nonsolid forming solvated ions from the reactants are certainly balanced and considered within the activity coefficient models (see the influence of the ionic strength in section 2.2.3).

3 The influence of mixing – Method development using the confined impinging jet mixer as a benchmark apparatus

Fast precipitation reactions are often influenced strongly by mixing, which masks the true kinetics of particle formation inherently. This leads to the point that resulting PSDs are hard to predict, especially in such a specific mixing-dominated regime. The capability of confined impinging jet mixers to adjust mixing times below the solid formation time enables to create process states that are not influenced by mixing. Thus, access to the fundamental kinetics of the precipitating system is given. Moreover, well-defined an increasing mixing influence can be adjusted by the flow condition which distinguishes CIJMs from other precipitation apparatuses such as stirred tank reactors.

This chapter strives to establish a fundamental understanding of the interplay of micro- and mesoscale, respectively, mixing and particle formation, during precipitation. Newly developed and enhanced methods which play a key element in considering the mixing phenomena during particle formation are developed. A CIJM is chosen to be an appropriate benchmark apparatus as it will be worked out in chapter 0. The methodic advancements of this chapter read as follows

- Introduction of experimental and numeric methods and setups that will be used.
- Generation of a phenomenological understanding of mechanisms occurring in CIJMs. An experimental database is created which allows for the validation of numerical simulations.
- The setup of a suitable CFD model that is used as the starting point to gain insight into transient mixing processes and to measure inaccessible state

variables dominating the solid formation (coarse-graining approach) experimentally.

- A method-development section followed by a model comparison with each other and with models provided in literature.

3.1 Experimental Setup

Figure 3.1 presents the CIJMs in the T- and Y-configuration investigated. The mixers consist basically of the following: A main duct, which is called the mixing chamber in the following. Two diminution sections in the entrance pipes lead into jet sections where fluids are accelerated towards the impinging point. The high entrance impulse is desired to generate high energy dissipation rates and, thus, high mixing intensities. The mixing geometries regarded are also used by Kucher et al. [21] and Kügler et al. [3] for their basic experimental examinations on precipitation kinetics. Geometric details can be extracted from Figure 3.1.

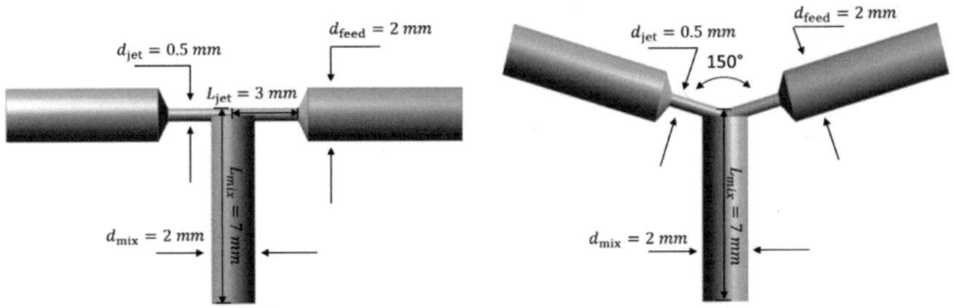

Figure 3.1: Confined impinging jet mixers investigated. Left: T-mixer ($\alpha = 180°$), right: Y-mixer ($\alpha = 150°$).

Experimental investigations to characterize the flow mechanics in CIJMs by laser-induced fluorescence (LIF) and particle image velocimetry (PIV) and the precipitation of the model system $BaSO_4$ were realized with three different experimental setups. Either two gear pumps (*Ismatec MCP-Z Process* with a *Z 130* pistonhead, see Figure 3.2, right), where the flow rate was controlled by two magnetic inductive flowmeters (*Krohne Optiflux 5300*), or two different syringe pumps, each with two pistons, were used (Figure 3.2, left).

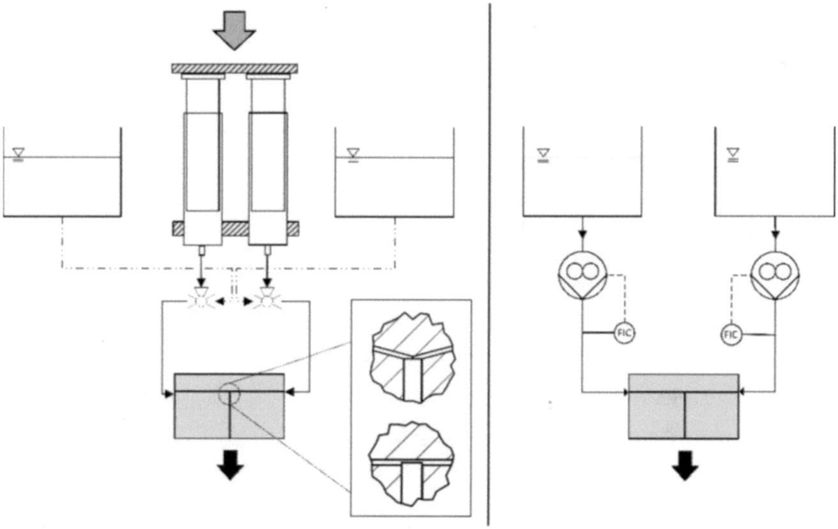

Figure 3.2: Two experimental setup types used; left: Batch operations using a syringe pump; the dotted ducts serve to refill the pistons from the reactant reservoir; right: Continuous gear pump setup controlled by two electromagnetic flowmeters.

A commercially available syringe pump (*Chemics Nexus 6000*, total liquid capacity $= 120\ ml$) was used to provide for the reactant volume flow for low flow rates. Additionally, a self-constructed piston pump (total liquid capacity $= 300\ ml$), integrated into a universal testing machine (*Zwick Roell*), that can handle and measure pressure drops up to $30\ bar$ with a piezoelectric pressure pickup, was used for high flow rates ($\mathrm{Re}_{\mathrm{mix}} \geq 3000$). The pressure drop was measured with either the pressure pickup mentioned or, for lower flow rates, a differential pressure manometer. This was built in just before one of the two inlets in the mixer and open to the environment via the low pressure duct.

3.1.1 Laser-spectroscopic flow field investigations

Laser-induced fluorescence and PIV measurements to study the flow dynamics of CIJMs are executed with an NdYAG Laser (*Dantec Dynamics*, wavelength $532\ nm$, frequency $15\ Hz$, energy $32\ mJ$, pulse-width $3-5\ ns$). Thereby, transparent PMMA CIJMs (tolerance of $\pm 0.01\ mm$) which are manufactured at the Institute of Micro Process Engineering at Karlsruhe Institute of Technology are investigated. A *HiSense MKII-CCD* camera (1344 x 1024 pixels, pixel size = $6.45\ \mu m$) with a macro prime lens (*AF Micro NIKKOR*, $= 60\ mm,\ 1{:}\,2.8\ D$) was used as a detector unit. Thereby, a local resolution of about 6 µm per pixel was achieved. This implies a reproduction scale of $1{:}\,1$. During the LIF investigation, the camera was equipped with a filter which absorbs in the wavelength of the laser light and which is transmissible for the fluorescence light.

The laser light is modulated with two plano-convex cylinder lenses ($f = 40\ mm,\ f = 60\ mm$) into a thin focal plane that allows only small slices ($\sim 200\ \mu m$ thickness) of the mixing geometry to be illuminated. The fluorescence component Rhodamine B was used for the LIF measurements. It absorbs in the wavelength range from $460 - 590\ nm$ (absorption maximum at $550\ nm$) and emits at a maximum of 590 nm [41]. Polyamide-Particles (*Dantec Dynamics*, monomodal with $L^P_{50,0} = 5\ \mu m,\ \rho^S_{PA} = 1140\ kg/m^3$) produced especially for PIV applications are used. Thereby, a small amount of about $0.5\ g/l$ is adequate to generate a high seeding density without changing the fluid behavior.

3.1.2 Precipitation experiments

Experimental investigations are carried out regarding the solid formation of the sparingly soluble salt $BaSO_4$ as a mixing-sensitive model system (see e.g. Schwarzer and Peukert [42]). Thereby, the precipitation in CIJMs is executed by mixing of the well soluble reactants $BaCl_2$ and Na_2SO_4 with the setups presented in Figure 3.2 at standard conditions ($\vartheta = 25\ °C,\ \mathrm{p} = 1\ bar$). The stainless steel mixers used are shown in Figure 3.1. The ionic reaction proceeds as illustrated in the following reaction equation.

$$BaCl_2(aq) + Na_2SO_4(aq) \rightarrow BaSO_4 \downarrow + 2\ NaCl(aq)$$

The solubility product K_{sp} of $BaSO_4$ is taken from Monnin [28] ($K_{sp,BaSO_4} = 9.98 \cdot 10^{-11}\ mol^2/l^2$). The practical activity coefficients $\gamma^r_{\tilde{m}_i} = \tilde{a}_i\ /(\tilde{m}_i/\tilde{m}_i{}^0)$ are calculated consonant with (see 2.2.3). In addition to the consideration of binary and ternary interaction parameters, the formation of the ion pair complex $[BaSO_4]_{aq}$ is included in the activity coefficient calculations (equilibrium constant used for the complex formation: $K_{IP} = 10^{-2.26}$ [29], see eq. 2.30). A nominal supersaturation of $S_{a,\mathrm{nom}} = 1000$ and $\mathrm{R_c} = 5$ is chosen for most of the experiments and the simulations presented. Thereby, certain assumptions explained in detail in section 2.4, for example, the electrostatic stabilization of the primary particles for $\mathrm{R_c} \geq 5$, are considered. Table 3.1 gives an overview of the initial ion concentrations that are chosen as a boundary condition at the inlets of the mixers. These concentrations refer to the nominal supersaturation of $S_{a,\mathrm{nom}} = 1000$ and the free lattice ion ratio $\mathrm{R_c} = 5$ in the case of instantaneous mixing and a mixing ratio of 1:1.

Table 3.1: Initial concentration for a nominal supersaturation of $S_{a,\mathrm{nom}} = 1000$ and $\mathrm{R_c} = 5$ in the mixing zone, calculated with the Pitzer approach. Due to the dilution in the mixing zone for a moment ratio $\mathrm{M_R} = 1$ of the feed-streams (small density differences are not considered here), the concentrations listed are twice the nominal values.

Initial molar concentration ($S_{a,\mathrm{nom}} = 1000$, $\mathrm{R_c} = 5$)	$(mol \cdot l^{-1})$
$\tilde{c}_{Ba^{2+},init}$	0.5803
$\tilde{c}_{Cl^{-},init}$	1.1606
$\tilde{c}_{SO_4^{2-},init}$	0.14465
$\tilde{c}_{Na^{+},init}$	0.28930

3.1.3 Characterization of precipitated particles

Particle sizes of the crystals suspended in their mother liquor are measured offline with static and dynamic light scattering technics. Thereby, the particle size range to be observed is decisive depending on which measurement device is used. Particles in the colloidal size range from 5 nm to 1 µm are measured based on the dynamic light scattering principle with a *Zetasizer Nano ZS* (*Malvern*) at a laser wavelength of $633\,nm$. Particles between $100\,nm$ to $30\,\mu m$ are measured additionally with a *Mastersizer 3000E* (*Malvern*) in a flow cell using static light scattering ($\lambda_{\mathrm{Laser}} = 633$ nm). The fundamental functionality is described in detail in Schlomach [43].

The SEM images are generated with either a *Leo 1530* (*Zeiss*, acceleration voltage $5\ kV$, using a secondary electron and an in-lens detector) or an ESEM *Quanta 650 FEG* (*FEI*) (acceleration voltage $15\ kV$). In a first step, particles are centrifuged ($20\ min$ at $1800\ g$) and dried for $24\ h$ at $80\ °C$. Subsequently, 2 to $3\ nm$ of platinum are sputtered on the sample surface to increase the electron conductivity.

3.2 Numerical setup and turbulence

Numerical investigations presented in the sections 0 and 3.4 are carried out by using the commercial solver *ANSYS Fluent 15.0* on two high-performance computing servers with *Intel Xeon Ivy Bridge* processors (16 Cores, 3.0 GHz, 128 GB RAM, each) and on the high-performance cluster *HP XC3000* (multicore calculations up to 128 cores) of Karlsruhe Institute of Technology. Turbulence is modeled with a detached eddy simulation-shear stress transport (DES-SST) k-ω approach. Thereby, a blend of large eddy simulation (LES) for energy rich vortices combined with a cost-efficient transient SST-k-ω two-equation formulation is used to model the highly transient flow field. The SST-k-ω approach, based on the unsteady averaged Navier-Stokes equations, is applied in the filtered subrange. Under the simplifying assumption of an incompressible fluid ($\rho = const.$), the general continuity eq. 3.1 (left) can be simplified to the second expression in eq. 3.1. The momentum equation for a Newtonian fluid $(\eta \neq f(\tau))$, describing the momentum change in a fluid element, reads as eq. 3.2, where $\vec{k}$ represents the body forces, for example, gravitation, ∇p the pressure gradient and $(\eta \cdot \Delta\vec{u})$ the contribution due to friction [44].

$$\frac{\partial \rho}{\partial t} + \nabla \cdot (\rho \vec{u}) = 0 \qquad resp. \quad \nabla \vec{u} = 0 \tag{3.1}$$

$$\rho\left(\frac{\partial \vec{u}}{\partial t} + (\vec{u} \cdot \nabla)\vec{u}\right) = \vec{k} - \nabla p + \eta \cdot \Delta\vec{u} \tag{3.2}$$

The *SST-k-ω* approach is solved in the spatial filtered LES-like sub-grid-scale, using the unsteady averaged continuity and Reynolds-averaged Navier-Stokes equations (URANS) [45, 46] (eq. 3.4 is the commonly used short form with

$i, j = x, y, z$). Detailed description of the full 3D set of equations are given in Oertel et al. [44])

$$\nabla \vec{\bar{u}} = 0 \tag{3.3}$$

$$\rho \frac{\partial \bar{u}_i}{\partial t} + \frac{\partial}{\partial x_j}\left(\rho \bar{u}_i \bar{u}_j\right) = -\frac{\partial \bar{p}}{\partial x_i} + \eta \frac{\partial}{\partial x_j}\left(\frac{\partial \bar{u}_i}{\partial x_j} + \frac{\partial \bar{u}_j}{\partial x_i}\right) - \frac{\partial}{\partial x_j}\left(\rho \overline{u_i' u_j'}\right) \tag{3.4}$$

The switch between RANS and LES is achieved by comparing the minimum of the turbulent length scale L_t in both modes.

$$L_t = \min\left(L_{t,\mathrm{RANS}}, L_{t,\mathrm{DES}}\right) = \min\left(\frac{\sqrt{k}}{C_\mu \omega}, C_{\mathrm{DES}} \Delta_{\max}\right) \tag{3.5}$$

Thereby, the empirical constants are $C_{\mathrm{DES}} = 0.78$ [47] and $C_\mu = 0.09$. $\Delta_{\max} = \max(\Delta x, \Delta y, \Delta z)$ is the maximal local resolution of the grid (grid spacing). The SST-k-ω transport equations are introduced for the closure of the Reynolds stress tensor $\rho \overline{u_i' u_j'}$ (eq. 3.6) in eq. 3.4.

$$-\rho \overline{u_i' u_j'} = \eta_t \left(\frac{\partial \bar{u}_i}{\partial x_j} + \frac{\partial \bar{u}_j}{\partial x_i}\right) - \frac{2}{3} \rho \delta_{i,j} k \tag{3.6}$$

Equation 3.7 shows the expression of the turbulent eddy viscosity for the k-ε and k-ω two-equation RANS turbulence approach.

$$\eta_t = \rho C_\mu \frac{k^2}{\varepsilon} = \rho \frac{k}{\omega} \tag{3.7}$$

The SST model equations used consists of a blend of the k-ε model for the bulk zones and the k-ω model for the near wall treatment due to the far simpler and more stable wall function of the k-ω model [48]. The additional equations 3.8 and 3.9 must be solved.

$$\frac{\partial}{\partial t}(\rho k) + \frac{\partial}{\partial x_i}(\rho k u_i) = \frac{\partial}{\partial x_j}\left(D_k \frac{\partial k}{\partial x_j}\right) + P_k - Y_k \tag{3.8}$$

$$\frac{\partial}{\partial t}(\rho \omega) + \frac{\partial}{\partial x_i}(\rho \omega u_i) = \frac{\partial}{\partial x_j}\left(D_\omega \frac{\partial \omega}{\partial x_j}\right) + P_\omega - Y_\omega \tag{3.9}$$

Where P_k is the generation of turbulent kinetic energy due to mean velocity gradients, P_ω is the generation of ω (characteristic frequency of energy-dissipating vortices), and Y_K and Y_ω are the dissipation of k, respectively, ω. D_k and $D_\omega$represent the effective diffusivity of k and ω. In the SST-k-ω approach [48], the blending between both models (k-ε and k-ω) is achieved by eq. 3.10.

$$\eta_t = \rho \frac{a_1 \cdot k}{max(a_1 \cdot \omega,\ F_2 \cdot S_r)} \qquad 3.10$$

Where F_2 is one of the SST blending functions [49], S_r is the strain rate magnitude and a_1 is an empirical constant. The calculation of single parameters and the calibration of empirical constants are extracted from Menter at el. [48].

The dissipation term for turbulent kinetic energy Y_K in eq. 3.8 in the DES-SST-k-ω model is modified by the length scale $L_{t,\mathrm{DDES}}$. The delayed-DES (DDES) shielding function (recommended by ANSYS) was chosen for the simulation. This shielding function is necessary to prevent the model from "grid induced separation (GIS)" while changing from the DES limiter to the boundary layers [50]. The DDES shielding function is newly calibrated. Nevertheless, it works in the same way as the blending functions in the SST-k-ω model.

$$Y_K = \frac{\rho\sqrt{k^3}}{L_{t,\mathrm{DDES}}} \qquad 3.11$$

Consonant with Gritskevich et al. [49], the DDES length scale is evaluated by

$$L_{t,\mathrm{DDES}} = L_{t,\mathrm{RANS}} - f_d \cdot max(0, L_{t,\mathrm{RANS}} - L_{t,\mathrm{LES}}) \qquad 3.12,$$

with f_d as the empirical blending function. The following parameter set was chosen for the simulation.

Table 3.2: Simulation parameters used for the STAR numeric measurements, presented in section 3.4.3

Feature	Approach selected
mode of calculation	transient
simulation approach	species model/single phase (liquid)
turbulence model	DES-SST k-ω
pressure-velocity coupling	simple
discretization	higher-order discretization (third-order MUSCL, bounded central differencing)
pressure discretization	PRESTO!
Lagrangian particle tracking	massless tracer particles

3.3 Fluid mechanics and precipitation in CIJMs

In comparison to other possible precipitation apparatus the CIJM is a highly efficient type of apparatus that can mix reactive fluids below $1\ ms$ and that, therefore, can decouple mixing and fast reactions. It is well-known that a characteristic plateau in the process performance can be found for increasing flowrates in CIJMs where a reaction or a particle formation step is no longer influenced by mixing. A region of turbulent high-energy dissipation, the constraint for all fluid elements to pass through this highly turbulent zone, and an oscillating mixing plane that amplifies mixing strongly have been identified as the key elements for the high efficiency of CIJMs [52, 53, 54, 55]. The CIJMs can cover a high-energy dissipation range ($10^{-1} - 10^{7}$ W/kg) far beyond the values that can be reached in ST reactors ($\sim\ 0.1\text{–}100\ W/kg$) [56, 55]. Due to this extensive range of mixing intensities and the unique ability of this apparatus to decouple mixing from particle formation which allows the study solely of the primary processes, the CIJM is considered as an important benchmark apparatus. Moreover, reproducibility is another important advantage.

Literature provides a bundle of experimental and simulative CFD-based publications on CIJMs. Short summaries on these works will be integrated in the following sections.

3.3.1 Experimental section

This section presents experimental findings achieved in CIJMs. Thereby, the focus is placed, on the one hand, on fundamental insights into the operating procedures of CIJMs, flow regimes and the transient flow dynamics in such devices. On the other hand, an experimental basis for the validation of CFD simulations executed in sections 3.4.2 and 3.4.3 is created. Important data, such as the effective pressure drop in the mixing zone which correlates to the energy dissipation rate as the dominant influencing parameter for (micro-) mixing and the study on the velocity of the segregation depletion of two unmixed feed-streams, are regarded. The Reynolds numbers used in this section ($i = \mathrm{mix}, \mathrm{jet}$) are defined as equation 3.13.

$$\mathrm{Re}_i = \frac{\bar{u}_i \cdot d_i}{\nu_{\mathrm{kin}}} \quad (3.13)$$

3.3.1.1 Pressure drop

The pressure drop is one sensitive operating parameter. On the one hand, cumulative loss of pressure with increasing flow rate in the entire apparatus leads to higher pumping costs, on the other hand, increasing effective pressure drop in the mixing zone improves the mixing efficiency due to a higher insertion of energy into the confined impinging zone (Figure 3.3, region marked grey). The effective pressure drops Δp_{mix} presented were obtained by subtracting all circumstantial pressure drops of the experimental setup caused by the hoses, flanges and bends downstream from the manometer to the CIJM with correlations given in literature [57].

An almost identical pressure drop is measured when Δp_{mix} for the T- and Y-mixer in Figure 3.3, right, are compared. A quadratic dependency of Δp_{mix} for $\mathrm{Re}_{\mathrm{mix}} \geq 1000$, indicating a proportionality of the pressure drop with $\bar{u}_{\mathrm{mix}}^2$, is observed. Pressure drop data presented will be used in the following for either the flow regime classification (chapter 3.3.1.4), the validation of CFD calculations (chapter 3.3.2.2) or the calculation of mean energy dissipation rates (chapter 3.4.3.7).

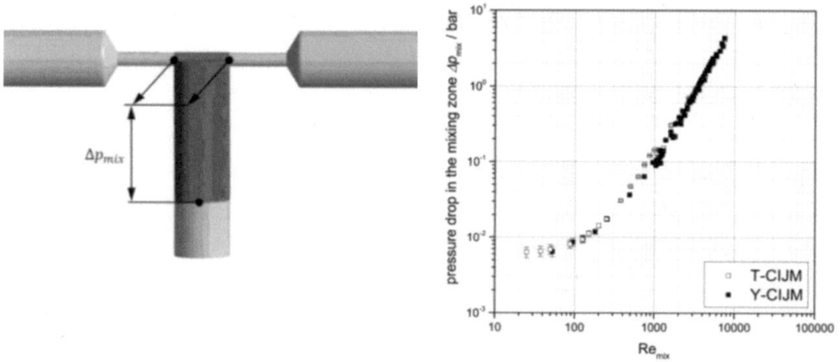

Figure 3.3: Effective pressure drop Δp_{mix} in the marked mixing zone for the Y- and T-CIJM extracted from experiments using a differential pressure manometer.

3.3.1.2 Transient flow dynamics in CIJMs

This section concentrates on investigations for obtaining a phenomenological idea of transient, turbulent mechanisms occurring in CIJMs. Data are visualized by LIF measurements for turbulent flow rates ($\mathrm{Re}_{\mathrm{mix}} \geq 1000$). Lower flow rate behavior is not investigated laser-spectroscopically within this thesis. Interesting works that handle such conditions are published by Santos and Sultan [58] and Sultan et al. [59].

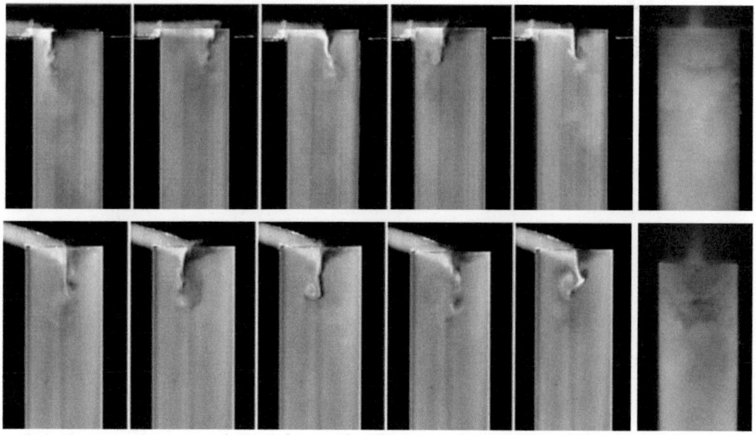

Figure 3.4: Consecutive snapshots of specific flow conditions and mixing in T (top)- and Y (bottom)-CIJMs at $\mathrm{Re}_{\mathrm{mix}} = 1600$. Right: 90° rotated view for the two CIJMs.

Figure 3.4 shows consecutive snapshots ($\Delta t = 200\ ms$) of the transient flow patterns visualized with LIF. The experimental procedure is explained in 3.1.1. Highly transient flow-field characteristics are induced by the two impinging jets in the T- and the Y-mixer. In order to complete the overall 3D picture of the fluctuation in the mixing zone, a rotation of the view plane of 90° (Figure 3.4, right) reveals that the central contact zone is a transient wavy structure of high gradients in turbulence, velocity, component distribution and pressure, as indicated in Figure 3.5. The mixing structure collapses downstream at a length of $\sim d_{\text{mix}}$ induced by intensive eddy shedding at its flattening tail. A transition to normal pipe flow is observed, which exposes a subsiding sinusoidal pulsation from the transient vortex plane. Furthermore, it can be stated that the Y-mixer has a pinned central plane, whereas the one of the T-mixer is in a metastable state and moves in a disorganized way from one jet side to the other. Assuming fast (quasi-instantaneous) solid formation processes, the risk is strongly enhanced, especially for industrial apparatuses, that one of the jet outlets may get blocked and the process collapses.

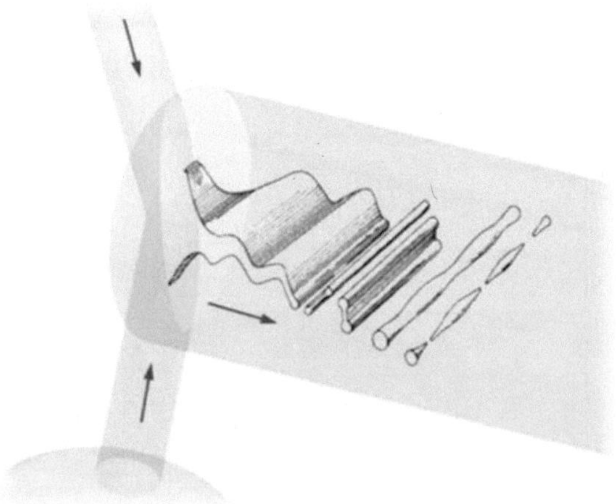

Figure 3.5: Conception of the oscillating mixing structure in the Y-CIJM. (Graphic adaption of the wavy structure from Dombrowski and Johns [60])

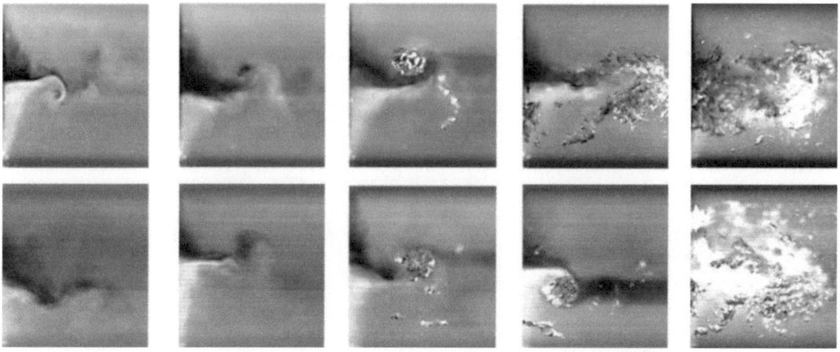

Figure 3.6: LIF Snapshots of the central observation plane at different flow rates in the Y-mixer (above) and T-mixer (below). From left to right: $\mathrm{Re}_{\mathrm{mix}}$ *= 1600, 2650, 3700, 4750, 5800.*

An overview of snapshots in Figure 3.6, concentrating on the head of the mixing chamber, shows for increasing flow rates (from left to right) that, in addition to an intensified eddy shedding achieved, further regimes, such as a vortex cavitation regime ($\mathrm{Re}_{\mathrm{mix}} \geq 3500$) with a transition to a cloud cavitation regime ($\mathrm{Re}_{\mathrm{mix}} \geq 5000$), are run through. It is found that the onset of these processes depends strongly on the counter pressure generated by the discharge of the mixer and cannot be expressed in a generalized way by a $\mathrm{Re}_{\mathrm{mix}}$ number. Details on such phenomena and their driving force and the ability of CFD to predict the cavitation behavior is discussed in the appendix (8.2.4) and in Metzger and Kind [54]. However, readers should keep in mind that cavitation, in addition, amplifies mixing strongly for high flow rates.

3.3.1.3 Depletion of segregation

One possibility to get spatially resolved information on the progress of mixing by experiments is the advancement of LIF frames presented in the previous section. Equation 3.14 represents the definition of the segregation index proposed by Danckwerts [61]. This approach quantifies the progress of mixing with the help of the standard deviation of concentrations c (or grey tones) of each pixel value to the mean of the global volume. The intensity of segregation, per definition, is normalized by the variance of the mean ${\sigma_0}^2$ to scale from 1 (completely segregated) to 0 (perfect mixing).

$$I_S = \frac{\sigma^2}{\sigma_0^2} = \frac{\overline{(c-\bar{c})^2}}{\bar{c}(1-\bar{c})} \qquad 3.14$$

The conversion of grey tones to I_S applied to the LIF images (Figure 3.4) leads to the images presented in Figure 3.7, wherein the white regions indicate segregated zones and black zones refer to a "mixed" system.

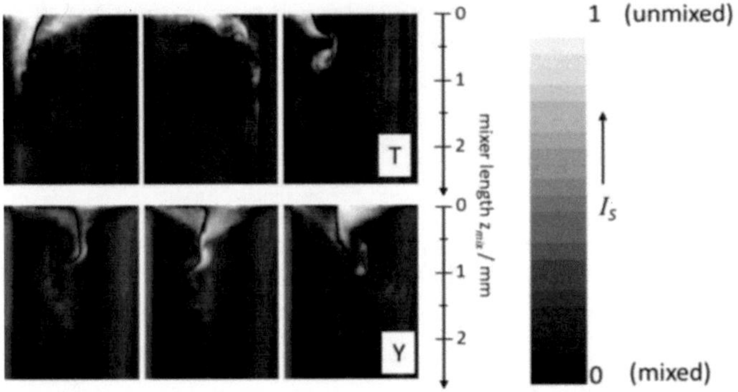

Figure 3.7: Transposed grey scale photos into segregation scale pictures: Top: T-mixer; Bottom: Y-mixer at $\mathrm{Re_{mix}} = 1600$.

It can be observed that the wavy vortex plane which is extended into the mixing geometry is the important zone of interest. A thin interfacial plane between the two colliding fluid streams is mixed directly and can be visualized as a black separating line within the light grey impinging zone. The final concentration gradient reduction step is reached by the transient fluttering and the breakup of the vortex plane.

Figure 3.8 shows the radially averaged axial progression of the segregation level in the mixing zone for the T- and Y-mixer at different flow rates at three instances each. An averaging of about 120 single LIF frames per flow rate was executed. This averaging suppresses the transient fluctuations and enables a representative comparison of the different flow rates. It is obvious that the characteristic region of intensive mixing is 2– 3 mm long and correlates well with the length of the mixing structure observed.

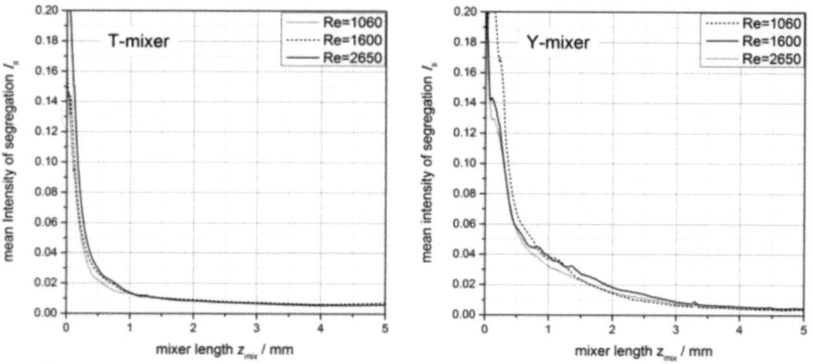

Figure 3.8: Radially averaged intensity of segregation evolution in the T- and Y-mixer for three different $\mathrm{Re}_{\mathrm{mix}}$ *numbers.*

By defining a point of mixture when segregation is decayed to 99 % ($I_s = 0.01$), the core region of mixing in the T- mixer is found to be $L_{\mathrm{mix},T} \approx d_h$. The one of the Y-mixer is slightly elongated due to an angled inflow ($L_{\mathrm{mix},Y} \approx 1.5 \cdot d_h$). Comparable results of the effective length of the mixing zone for turbulent Reynolds numbers are reported in literature. An almost constant $L_{\mathrm{mix}}/d_h \approx 1.5$ is found in the T-shaped mixing head by Teixeira et al. [62]. Considering the sensitivity of the measurement method, I_s decays almost completely along the mixing zone regarded. An increase of the flow rate does not change the local dependence of the segregation level significantly. This means that mixing is dominated by inertial-convective phenomena with the time constant inversely proportional to the average velocity $\bar{u}_{\mathrm{mix}}$. Since the time constant of turbulent mixing on the mesoscale observed experimentally is proportional to $\varepsilon^{-1/3}$ [34], it can be concluded that $\bar{u}_{\mathrm{mix}} \propto \varepsilon^{1/3}$

3.3.1.4 Flow regimes in CIJMs

Specific flow regimes for CIJMs are classified in the literature. As reported by Bothe et al. [63], Santos and Sultan [58] and Sultan et al. [59], the laminar regime for the rectangular T-shaped microchannels investigated can be subdivided in a "segregated or stratified" flow regime for very low Re numbers ($\sim \mathrm{Re}_{\mathrm{mix}} \leq 50$) and a "vortex flow" regime where two vortices with a counter rotating axis are built which do not effectively exchange matter ($\sim\mathrm{Re}_{\mathrm{mix}} \leq 100 - 300$). Since the true transition values depend strongly on the mixing geometry investigated, only a rough quantitative classification is possible from the literature. However, it is interesting that the regime transitions reported scatter relatively little for varying mixing chamber diameters from $d_{\mathrm{mix}} = 0.4\ to\ 6\ mm$.

An "engulfment flow" regime is reported when increasing the flow rate into a transition state. Santos and Sultan [58] give a broad overview of the transition Reynolds numbers measured by different authors. Thereby, the Reynolds range for the transition observed lies between ($\sim 130 \leq \mathrm{Re}_{\mathrm{mix}} \leq 300$). A fully turbulent regime is observed ($\sim\mathrm{Re}_{\mathrm{mix}} > 500$) for a further increase in flow rate.

Investigations are made regarding an adequate classification of the two mixers investigated and to get an insight into the dominant flow regimes. Pressure drop data are converted into a specific pressure loss coefficient. This coefficient ξ, calculated according to eq. 3.15 (see Figure 3.9) by considering the mean jet velocity $\bar{u}_{\mathrm{jet}}$ according to the VDI Heat Atlas [57], gives certain important information.

$$\xi = \Delta p_{\mathrm{mix}} \, / \left(\frac{\rho^L}{2} \cdot \bar{u}_{\mathrm{jet}}^2\right) \qquad 3.15$$

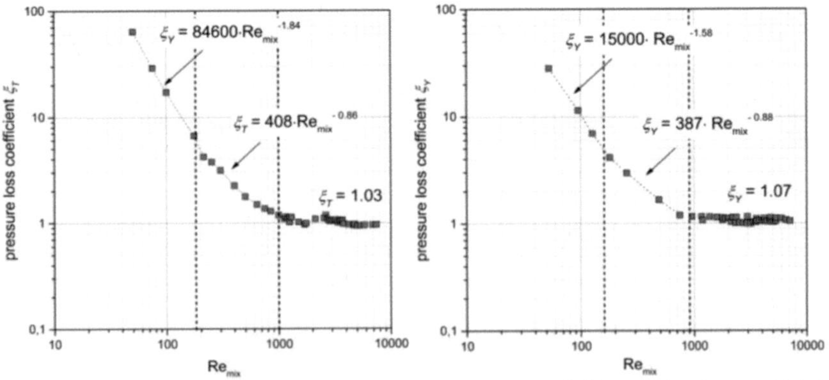

Figure 3.9: Pressure loss coefficients diagram for the T-mixer (left) and the Y-mixer (right), $d_{mix}/d_{jet} = 4$ *(geometries, see Figure 3.1) and a momentum ratio* $\mathrm{M_R} = 1$.

Both mixers show about the same ξ for $\mathrm{Re_{mix}} \geq 1000$ (i.e. $\xi_T = 1.03$ and $\xi_Y = 1.07$), which is, following the considerations of a Nikuradse diagram, supposed to represent the turbulent regime within the mixers. Lower flow rates lead to an increase of the pressure loss coefficient. In contrast to the Nikuradse diagram for tubes that involve a dependency of the laminar flow regime with $64/\mathrm{Re}$ and, thus, a slope of -1, differing slope values for the CIJMs were measured. Both mixers show a change in the linear (laminar) slope at a critical $\mathrm{Re_{mix}}$ number between 180 and 200, see Figure 3.9. The range in between is supposed to be the transition region, which is known in literature as the "engulfment flow" regime. A classification of different flow regimes is made considering the flow regimes observed and classified in literature and by evaluating pressure drop data (3.3.1.1) and LIF observations (3.3.1.2). The findings are in good accordance with the regimes found in literature. The different experimental observations allow a coherent picture which will be consolidated by additional findings subsequently in section 3.3.1.5 and in the numerical section 3.3.2.2. The CFD results confirming the non-cavitating flow regimes well are presented in the appendix (Figure 8.5).

Table 3.3: Classification into specific flow regimes in the CIJMs investigated according to experimental findings and considering the literature [58]. The mean energy dissipation (discussed in section 3.4.3.7) is added to the Re_{mix} numbers where a transition is observed.

Flow regime	$\mathrm{Re_{mix}}$	$\bar{\varepsilon}$ /(W·kg^{-1})
Laminar flow		
▪ Segregated flow	≤ 50	≤ 10
▪ Vortex flow	$50 - 200$	$10 - 220$
Transition flow		
▪ Engulfment flow (transition 1)	$200 - 1000$	$220 - 8700$
Turbulent flow		
▪ Chaotic engulfment flow	$1000 - 3500$	$8700 - 2.8 \cdot 10^5$
▪ Vortex cavitation (transition 2)	$3500 - 4500$	$2.8 \cdot 10^5 - 5.7 \cdot 10^5$
▪ Cloud cavitation	> 4500	$> 5.7 \cdot 10^5$

However, one should keep in mind that mixing in CIJMs may be not fully characterized by the $\mathrm{Re_{mix}}$ or $\mathrm{Re_{jet}}$ number, since the energy dissipation rate ε inserted into the mixing chamber contributes mainly to the mixing rate and is responsible for the vortex cascade that leads to the depletion of concentration gradients (see section 3.4.4).

3.3.1.5 Precipitation in CIJMs – the influence of mixing

The mixing step plays a crucial role in fast reacting systems, especially when the mixing and the consecutive step, such as chemical reaction or the particle formation due to a supersaturation buildup, occur on equivalent time scales. Incomplete mixing and inhomogeneous fluid zones lower the nominal rates of the specific processes significantly. Figure 3.10 shows the mean particle sizes $L_{50,0}^{P}$ (left) and $L_{50,3}^{P}$ (right) measured in the CIJMs (T- and Y-configuration) that will be discussed in detail during the sections in chapter 3. Firstly, one can identify that both mixer devices show rather similar results. It is well-known that a characteristic plateau in the process performance can be found for high flow rates in CIJMs where the reaction or particle formation step is no longer influenced by mixing [21, 64, 65]. This behavior, also found in Figure 3.10 for $\mathrm{Re}_{\mathrm{mix}} \geq 1000$, matches the starting point for a fully turbulent flow well (see Table 3.3).

The high turbulence generated in this type of apparatus enables the mixing of reactive fluids very fast (~ms, see details later in section 3.4.3.6).

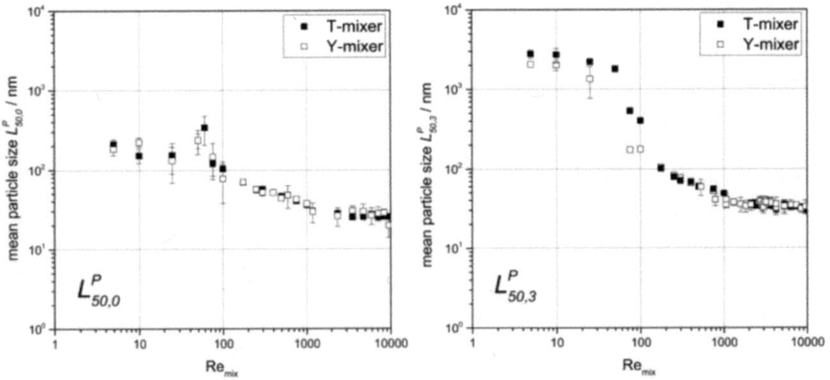

Figure 3.10: Mean particle diameter for the precipitation of $BaSO_4$ at $S_{a,\mathrm{nom}} = 1000$ and $\mathrm{R_c} = 5$ plotted over the flow intensity $\mathrm{Re}_{\mathrm{mix}}$: Left: $L_{50,0}^{P}$, right: $L_{50,3}^{P}$, mixers used are illustrated in Figure 3.1

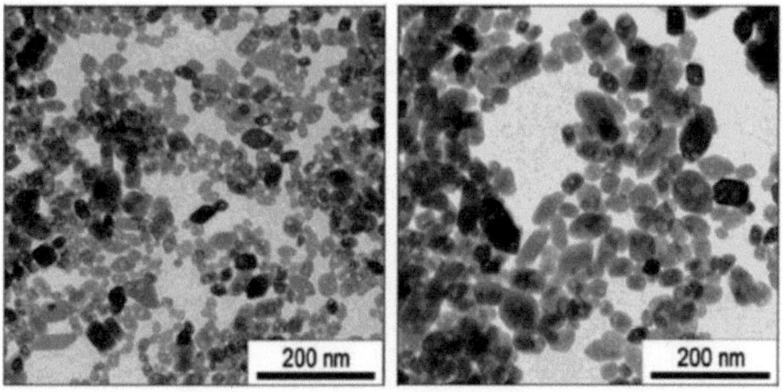

Figure 3.11: Transmission electron microscopy images of $BaSO_4$, precipitated in the mixing geometry presented at $S_{a,\text{nom}} = 1000$ and $\text{R}_\text{c} = 5$: left, at $\text{Re}_\text{mix} = 7130$; right, at $\text{Re}_\text{mix} = 535$, extracted from Kucher and Kind [4].

Thus, mixing and fast reaction are decoupled and the possibility exists to measure the "true" kinetic of particle formation which is not masked by mixing. An increasing but highly reproducible particle size is observed by decreasing flow rates entering the "engulfment flow" regime. A mixing phenomenon starts to superpose the solid formation process. Transmission electron microscopy images in Figure 3.11 give a very distinguishable, qualitative impression of the influence of mixing conditions on the size of the resulting particles.

A further decrease of the flow rates leads to laminar mixing conditions $\text{Re}_\text{mix} \leq 100$, where another significant change in particle size is observed. The mean particle sizes $L^P_{50,0}$ and $L^P_{50,3}$ which, so far, have been quite close to each other for the transient and turbulent regime, start deviating strongly. The dispersion of the size distribution increases distinctly. Precipitation under laminar mixing conditions will be discussed in section 3.4.3.5.

3.3.2 Numerical section and comparison to experiments

The numerical modeling of complex flow phenomena with CFD methods has attained increased interest in the last decades. Steadily improved models regarding numerics and improved turbulence modeling have led to an improved reproduction of reality.

Confined impinging jet reactors have been modeled frequently with a two-equation RANS approach, for example, the k-ε realizable or k-ε standard-model [66, 67, 68] and [18]. These approaches involve a strong averaging of the flow field. The influence of turbulent deviation is induced with a fluctuation term of the turbulent viscosity that is underestimated strongly, as experimental LIF and PIV publications [69] show. Comparing the simulation and experiment, workgroups of Icardi et al. [70] and [71], Gavi et al. [72] and Makowski and Badyga [73] showed that the simulation of flow conditions in CIJMs near to reality requires more complex turbulence modeling. These findings were also demonstrated by Schwertfirm and Manhart [74] in their comprehensive mixing scale Semi-DNS simulation in a rectangular T-mixer, associated with high computational load. Consequently, instead of steady CFD simulations in such mixers, transient dynamic calculations must be carried out. The flow field in CIJMs has a strong 3D nature. Santos et al. [75, 76] and Soleymani et al. [77] showed that 2D simulations are only able to provide reliable information in the laminar steady state regime for $\mathrm{Re_{mix}} \leq 200$. A 3D geometry has got to be modeled in CFD for higher flow rates.

This section introduces the numerical grid which will be used in chapter 3.4. Thereby, a grid independency study and an experimental validation of the transient flow field simulated is executed. The CFD will be used as an important instrument for the method development in sections 3.4.2 and 3.4.3.

3.3.2.1 Grid independency

The 3D, partly hybrid meshes were created for the numerical investigation of the T- and Y-CIJM presented. A gradient-refined tetrahedral mesh grid was assumed to be a good choice based on the high level of complexity at the inlet from the jet zone to the mixing zone and the need for a higher cell density than the rest of the geometry due to high local gradients in the current. Regions of minor interest are meshed coarser.

Table 3.4 gives an overview of the characteristic grid data of three different meshes that were tested for grid independency. In all these cases, the influence of the boundary layer of the wall was of particular importance. Except for mesh 1, the geometry surfaces are covered with five or ten prism-layers that are refined towards the wall to resolve the boundary layer.

Table 3.4: Data of different meshes tested of the T-mixer

T-mixer	Nodes	Elements	Wall prism layers	Refinement
Mesh 1	101 642	327 003	0	⇩
Mesh 2	244 697	500 587	5	
Mesh 3	1 124 778	1 802 722	10	

Transient simulations with sophisticated turbulence models, such as LES or DES simulations, induce high turbulent fluctuations into the flow field that hamper comparability of the meshes presented severely. Therefore, grid independency calculations are carried out with the k-ε realizable model in a stationary mode at comparatively high flow rates ($Re_{mix} = 4000$), where turbulent length scales are small. It is assumed that the findings are reasonably transferable to other types of turbulence modeling and certainly valid for lower flow rates. As an example, Figure 3.12 shows the data of the velocity field in the mixing zone of one central line 5 mm after the inlet. Whereas mesh 1 does not include refined prism cells near the wall, mesh 2 and mesh 3 show an improved resolution of the boundary layer in the mixing zone. This also affects the fluid mechanics in the bulk.

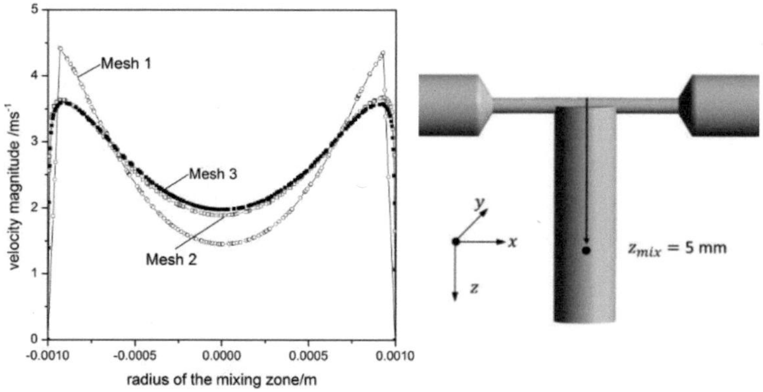

Figure 3.12: *Grid independency study of the meshes given in Table 3.4. The velocity magnitude of a line in the y-direction at $z_{mix} = 5\ mm$ and $Re_{mix} = 4000$ is observed.*

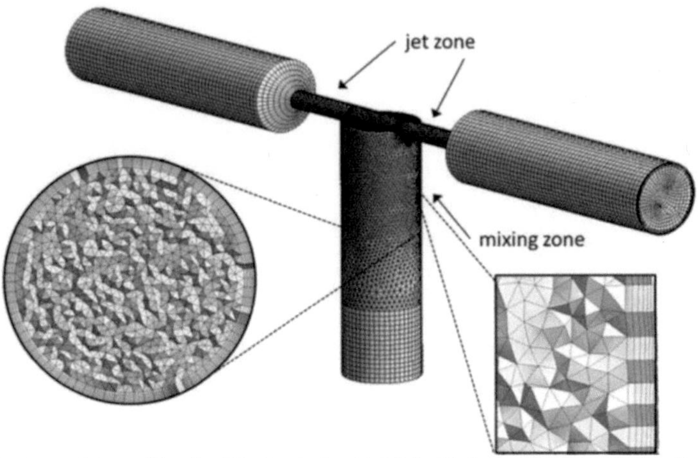

Figure 3.13: Mesh 2 used for the CFD simulation, T-CIJM with $d_{\mathrm{mix}} = 2\ mm$ and a jet diameter of $d_{\mathrm{jet}} = 0.5\ mm$ [54, 78].

Infinitesimal changes due to a distinct increase of elements from mesh 2 to mesh 3 are tolerated for the sake of low computational cost. Regarding the data generated, mesh 2 (see Figure 3.13), a hybrid mesh with a gradient-refined mixing zone and resolved boundary layer throughout the whole geometry, was taken as the basis for all the following calculations.

3.3.2.2 Validation of CFD simulations – comparison to experimental data

Exemplarily selected comparisons to experiments are presented in this section. One indicator of the transient CFD simulations which compares well with the experiments is the effective pressure drop in the mixing chamber (chapter 3.3.1.1). Figure 3.14 shows the experimental data of the T-mixer plotted versus those achieved in a transient DES-SST-k-ω simulation averaging the data for at least 5 s of process time. The pressure drop from CFD is calculated by measuring the mean total pressure of the system at the two inlet planes and in a plane $5\ mm$ downstream of the mixing chamber (see dark grey region in Figure 3.3, left), consonant with eq. 3.16.

$$\Delta p = \frac{p_{\mathrm{tot,inlet\,1}} + p_{\mathrm{tot,inlet\,2}}}{2} - p_{\mathrm{tot},z5} \qquad 3.16$$

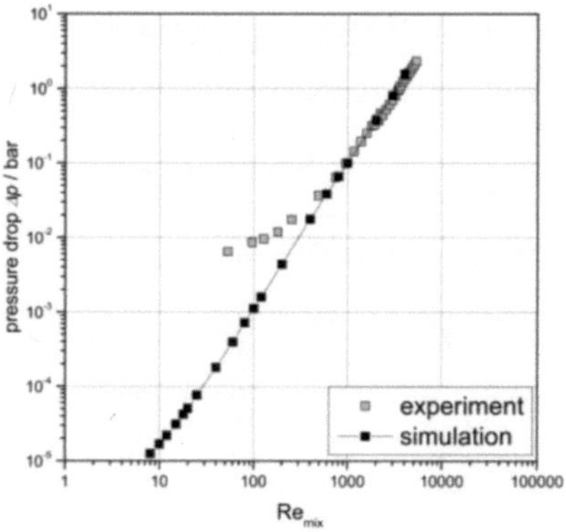

Figure 3.14: Comparison between the effective pressure drop measured in the T-mixer investigated and the data received within CFD.

Highly congruent results can be measured in the turbulent and most of the transition regime. Deviations between the simulation and experiments are found for $\mathrm{Re}_{\mathrm{mix}} < 300$. The reason for this deviation is not fully clarified. On the one hand, the accuracy of measurement and the accuracy of the flow rate which is experimentally adjusted decreases with decreasing $\mathrm{Re}_{\mathrm{mix}}$. On the other hand, defective values may arise from CFD since the turbulence model (SST-k-ω) used in zones of low turbulence is calibrated for fully turbulent flows. However, a good accordance is found especially for higher flow rates.

A PIV image acquisition is used regarding the dominant flow patterns and transient behavior of fluid mechanics in the mixer. A defined slice of the 3D vector field is extracted by using the thin focal plane of the PIV setup in polymethyl methacrylate (PMMA) mixers (constructed in a similar way to the geometries showed in Figure 3.1). Experimental results in the cylindrical geometry are generated with the refractive index-corrected system water-NH_4SCN (details are given in the appendix) using polyamide tracer particles, as described in chapter 3.1.1. Two exemplary snapshot frames taken of the velocity magnitude profile of PIV measurements at $\mathrm{Re}_{\mathrm{mix}} = 1030$ are compared to simulation results in the following.

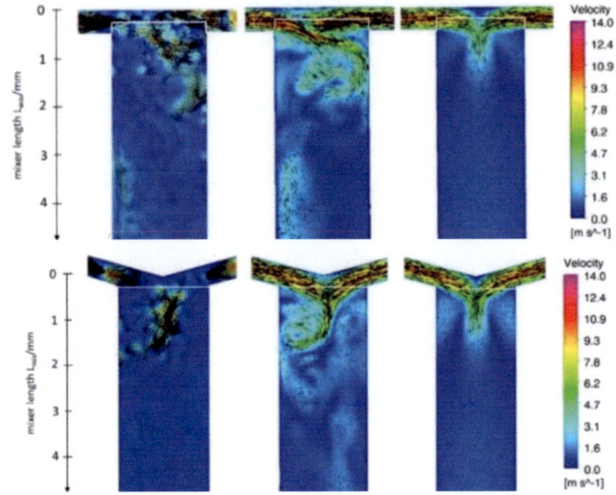

Figure 3.15: Comparison of the morphology of the velocity vector field in the T- and Y-CIJM at $\mathrm{Re_{mix}} = 1030$. *Left: PIV frame, middle: DES-SST-*k*-*ω*-turbulence model, right:* k*-*ε *Re normalization group (RNG) turbulence model.*

Analogical snapshots are compared due to the high transient dynamics in the mixing nozzles during the experiments and in DES. Nevertheless, it should be mentioned that a fitting snapshot can be found in the DES-CFD simulation for almost every particular flow situation in the experiment.

Obviously, the comparison of screenshots (Figure 3.15) demonstrates that the k-ε turbulence model is not adequate to describe transient flow conditions of confined impinging jets in such mixers accurately, which is not surprising. Thereby, a strong averaging effect is induced into the simulation which changes the highly transient behavior observed significantly. However, this transient nature of the flow field is a main impetus for mixing in such mixing nozzles. Nevertheless, the value of the velocity magnitude (eq. 3.17) averaged in the k-ε mode fits quite well to the data generated by PIV (see Figure 3.16).

$$u_{\mathrm{mean}} = \sqrt{u_x^2 + u_y^2 + u_z^2} \qquad 3.17$$

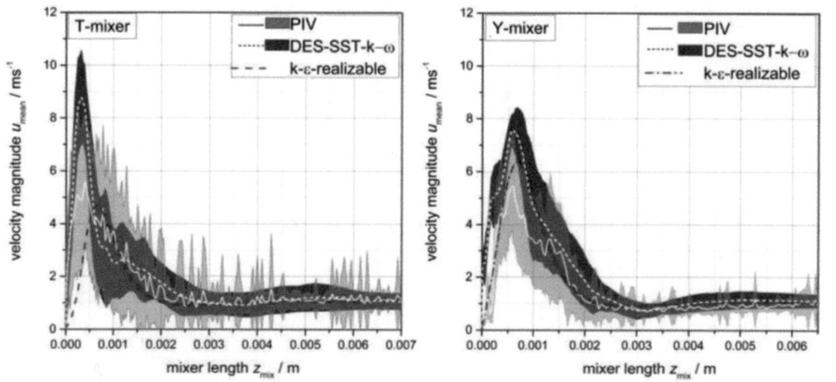

Figure 3.16: Time-averaged (~60 single frames) mean velocity at a line located in the center of the mixing zone in the flow direction. Grey zones indicate the fluctuation of velocity. The k-ε model does not show fluctuations.

The results are highly congruent concerning the location of maxima in velocity magnitude as well as the turbulent fluctuations in the mixing zone (grey zones in Figure 3.16) for the DES and the velocity profile measured. Based on the geometrical settings in the head of the mixing zone (transition from jet pipe $d_{\text{jet}} = 0.5\ mm$ to mixing pipe $d_{\text{mix}} = 2\ mm$), experimental PIV data extracted in this zone is slightly defective. The nonopaque edge (white line in Figure 3.15) leads to small errors in the cross-correlation algorithm, using the noncommercial *Matlab* package *PIVlab*. Therefore, the maximal values of velocity are slightly underestimated and a characteristic little buckle (Figure 3.16) at $z_{\text{mix}} \approx 0.25\ mm$ can be observed.

Nevertheless, it should be mentioned that the DES-SST-k-ω model is not a direct numerical simulation. Thus, a certain averaging effect, especially in the fluctuation, should be expected (Figure 3.16). Experiments at $\text{Re}_{\text{mix}} = 1030$ and DES simulations, shown in Figure 3.15, illustrate that the fully developed turbulence in the head of the mixing zone induces a sinusoidal flow pattern along the mixing zone. Thereby, the T-mixer uses the complete hydraulic diameter d_h of the mixing zone as an amplitude. In addition to the continuous vortex formation by screwing in the pined mixing structure, the Y-mixer tends towards a higher frequent fluctuation with smaller amplitude.

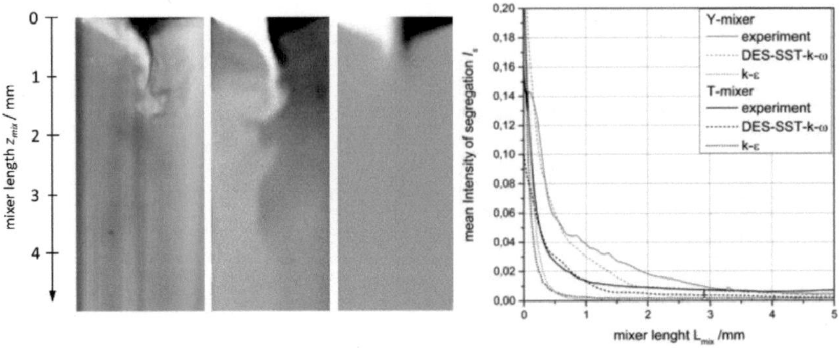

Figure 3.17: Left: Comparison of experiment and CFD using different turbulence models at $\mathrm{Re_{mix}} = 1600$*, frame 1: experiment, frame 2: DES-SST-k-ω model, frame 3:* k*-*ε *model; right: Radially averaged intensity of segregation evolution in the T- and Y-mixer for* $\mathrm{Re_{mix}} = 1600$*, comparison between experiment and simulation (averaged values of 60 to 120 frames).*

Turbulence is a highly stochastic effect, so that one specific frequency in either of the mixer types could not be extracted.

A first idea regarding a frequency range extracted for the T-mixer by Fourier Transformation will be given in the appendix. As has already been shown in section 3.3.1, the velocity field observed (PIV and CFD) also supports the conclusion that the zone of high turbulence, high velocities and high mixing efficiencies is restricted to the first $2\ mm$ (T-mixer) to $3\ mm$ (Y-mixer) of the mixing chamber in the turbulent regime investigated.

Figure 3.17 left, compares a snapshot of a LIF measurement (left) in the Y-mixer with simulated data of the mixing of two fluids (fluid 1 (fluorescent): white, fluid 2: black) using a detached eddy (mid) and a k-ε realizable approach (right). Thereby, the DES-SST-k-ω has been found to be well in accordance with experimental findings comparing radial and time-averaged intensity of segregation in the T- and Y-mixer (Figure 3.17, right). The k-ε-model, as a computationally cost-efficient and commonly used model in industrial application, overpredicts mixing in the turbulent case presented, Even for the modeling of mixing problems with CFD methods.

Conclusions

This section summarizes essential experimental and numerical results.

- An experimental basis for the understanding of complex flow and mixing procedures in CIJMs is generated.
- Effective pressure drop measurements, the indirect interpretation of mean particle sizes gained by precipitation and LIF studies allow a classification of certain flow regimes in the CIJMs investigated.
- Comparison between T- and Y-mixer only showed small differences. It can be stated that the main difference between the two devices is an enlarged mixing zone with a stronger pinned wavy mixing plane for the Y-mixer
- A grid-independent numerical mesh is presented which will be used for CFD calculations executed in chapter 3.
- Numerical and experimental results are compared to each other. The DES-SST-k-ω turbulence model is found to be a very suitable approach to predict the highly transient flow field appropriately.
- It is found that the averaging effect of the k-ε model underestimates mixing significantly.

3.4 Method development – coupling mixing and particle formation

The description of the mixing step and the consecutive buildup of supersaturation are determining for the chemical reaction on a molecular level and the subsequent solid formation. In addition to the intrinsic kinetics of nucleation and growth, characteristic attributes of the final product, such as the PSD achieved, depend strongly on the mixing characteristics of the particular apparatus and the mixing situation. When setting up a predictive model for precipitation processes, the consideration of such mixing influences is challenging, since multiple scales from molecular to micro-, meso- and macromixing phenomena must be covered. Three different methods are introduced in the following (see Figure 3.18) which differ strongly in complexity, their levels of detail and, therefore, differ strongly in computational load as well.

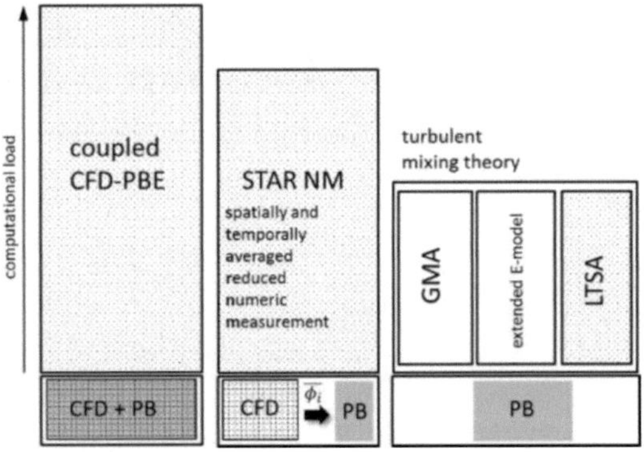

Figure 3.18: Scheme of approaches investigated in this chapter. One should differentiate between the three main methods (three columns presented).

The first method introduced (see section 3.4.2) is a fully coupled CFD-PBE approach. Results shown in section 3.4.1 and 3.4.2 are published in Metzger and Kind [30]. The second newly developed methodological approach consists of a decoupled PBE approach fed with mean state variables and residence time information, which are gained from CFD measurements. In the following, this coarse-graining approach is called "STAR NM" which stands for

Spatially and Temporally Averaged Reduced Numeric Measurement. Thereby, CFD is applied as a "measuring" instrument to get access, for instance, to mean residence time distributions and spatially and temporally averaged supersaturation values. Results achieved by this approach are published in Metzger and Kind [79].

A third type of approach with the lowest computational effort completely resigns on CFD methods and is based on turbulent mixing theory in which the progress of mixing is derived by multiple coupled ordinary differential equations (ODEs). Thereby, two approaches from literature (GMA = global mixing approach [42] and the extended E-model (= extended engulfment model) including mesomixing phenomena [80]) are applied for the CIJMs regarded. The limiting time scale approach (LTSA), which will be presented additionally, represents a newly proposed approach improvement which can tackle the scale-up of CIJMs. A detailed overview is published in Metzger and Kind [78].

3.4.1 The influence of "sub-grid" mixing within CFD

One central element is investigated to rate the accuracy of the CFD simulations upstream to the investigation of approaches announced. As reported in literature [81, 72] and [82], mixing on the grid scale may not describe the gradient-reduction process properly, especially for precipitation reactions where mixing on a molecular scale is of particular interest.

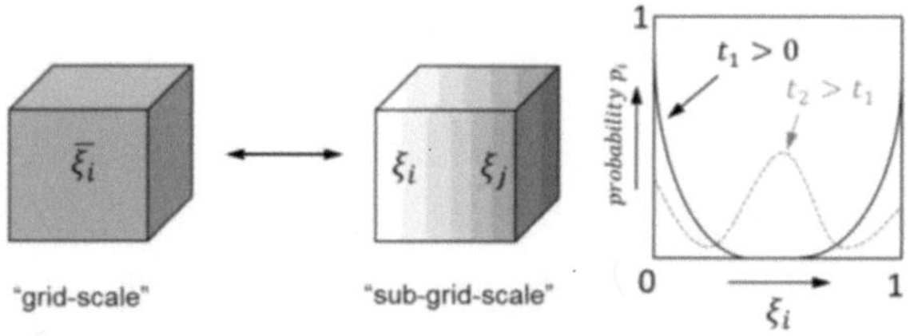

Figure 3.19: Basic idea of sub-grid models for micromixing within CFD.

Therefore, micromixing models have been established that describe sub-grid gradients (Figure 3.19) in a statistical way as probability density functions (PDFs) [81]. One model that has received increased attention in the last few years, since it can be nicely implemented into the CFD framework, is the DQMOM-interaction by exchange with the mean (IEM) approach, proposed by Marchisio and Fox [16]. Thereby, the classic IEM model (eq. 3.18), first proposed by Harada [83] and Costa and Trevissoi [84], assumes that a mixture mass fraction ξ_i of a component i exchanges mass with the local mean value of the cell $\bar{\xi}_i$ (see eq. 3.18). This work neglects the reaction rate r_i due to investigations solely of the mixing step.

$$\frac{d\xi_i}{dt} = k_m(\bar{\xi}_i - \xi_i) + r_i \qquad 3.18,$$

where k_m represents the rate constant of mixing in the IEM model that is defined as eq. 3.19.

$$k_m = C_\xi \frac{\varepsilon}{2k} \qquad 3.19,$$

k is the turbulent kinetic energy and ε the energy dissipation rate in the grid cell regarded. Literature has various formulations for the mixing constant C_ξ; some [85, 86, 87] and [88] choose $C_\xi = 2$. Ackroyd et al. [89] and Tsai and Fox [90] take a value of $C_\xi = 1.65$. It is decided to follow the work of Liu and Fox [18], who extracted C_ξ from a scalar energy spectrum for $\mathrm{Sc} = 1000$, depending on the turbulent Reynolds number $\mathrm{Re_T}$ (see eq. 3.20). Liu and Fox [18] have shown that the micromixing rate is over-predicted by taking a fixed value of $C_\xi = 2$ for laminar mixing at lower flow rates.

$$\mathrm{Re_T} = \frac{k}{(\varepsilon \cdot \nu_{\mathrm{kin}})^{0.5}} \qquad 3.20$$

Due to numerical problems in the boundary area caused by erratic values of their sixth-order polynomial fitting [18] (see dashed line, 3.11), their data are refitted for more stable conditions within the CFD calculations.

The refitted data using a Hill fit (black line in Figure 3.20) read as eq. 3.21.

$$C_\xi = C_{\xi,\max} \cdot \frac{Re_T^{n_1}}{a^{n_1} + Re_T^{n_1}} \qquad 3.21,$$

where $C_{\xi,\max} = 2.26638$, $a = 5.51115$ and $n_1 = 0.91242$.

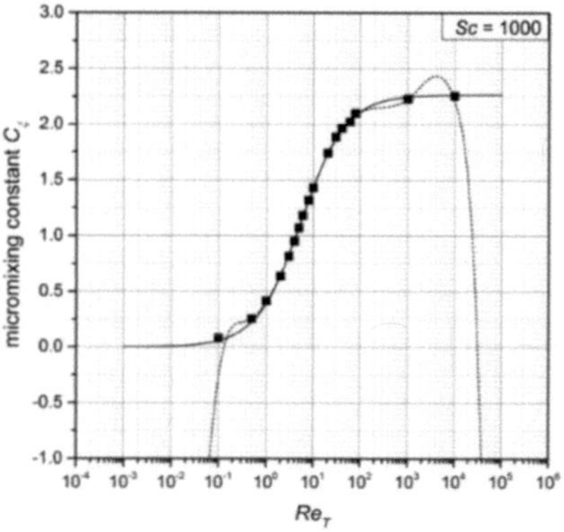

Figure 3.20: Dependence of C_ξ on the turbulent Reynolds number for $Sc = 1000$, according to Liu and Fox [18]. Dashed line: polynomial using the polynomial coefficients from Liu and Fox [18], black line: refitted data, eq. 3.21.

Marchisio and Fox [16] embedded the basic IEM micromixing approach into the DQMOM procedure that enables a PDF distribution to be approximated with the sum of N Dirac delta functions. This model is very convenient for CFD applications due to the low computational costs and the mathematically correct closure formulation of the first moments of the joint PDF approach [18]. The DQMOM-IEM model is implemented choosing the number of environments $N = 2$. Thereby, three additional scalar transport equations for the probability of environment 1 p_1 and the weighted probabilities s_1 ($= p_1\xi_1$) and s_2 ($= p_2\xi_2$) must be solved within CFD. The fourth scalar p_2 (probability of environment 2) can be calculated from the closure (eq. 3.25) that represents the zeroth moment of the DQMOM approach.

$$\frac{\partial \rho p_1}{\partial t} + \frac{\partial}{\partial x_i}(u_i \rho p_1) - \frac{\partial}{\partial x_i}\left(D_t \rho \frac{\partial p_1}{\partial x_i}\right) = 0 \qquad 3.22$$

$$\begin{aligned}\frac{\partial \rho s_1}{\partial t} + \frac{\partial}{\partial x_i}(u_i \rho s_1) - \frac{\partial}{\partial x_i}\left(D_t \rho \frac{\partial s_1}{\partial x_i}\right) &= k_m \rho p_1 p_2 (\xi_2 - \xi_1) \\ &+ \frac{D_t}{\xi_1 - \xi_2} \rho \left(p_1 |\nabla\langle\xi_1\rangle|^2 + p_2 |\nabla\langle\xi_2\rangle|^2\right)\end{aligned} \qquad 3.23$$

$$\begin{aligned}\frac{\partial \rho s_2}{\partial t} + \frac{\partial}{\partial x_i}(u_i \rho s_2) - \frac{\partial}{\partial x_i}\left(D_t \rho \frac{\partial s_2}{\partial x_i}\right) &= k_m \rho p_1 p_2 (\xi_1 - \xi_2) \\ &- \frac{D_t}{\xi_1 - \xi_2} \rho \left(p_1 |\nabla\langle\xi_1\rangle|^2 + p_2 |\nabla\langle\xi_2\rangle|^2\right)\end{aligned} \qquad 3.24$$

$$p_2 = 1 - p_1 \tag{3.25}$$

In addition to the convection term, the three transport equations involve the diffusion term considering turbulent diffusion and a source term on the right-hand side of the equations. The turbulent diffusion coefficient D_t is calculated with eq. 3.26, wherein $C_\mu = 0.09$ and $\mathrm{Sc_T} = 0.7$.

$$D_t = \frac{C_\mu}{\mathrm{Sc_T}} \frac{k^2}{\varepsilon} = \frac{1}{\mathrm{Sc_T}} \frac{k}{\omega} \tag{3.26}$$

The source term formulations are taken from the work of Liu and Fox [18] and Gavi et al. [72]. The scalar transport equations of the weighted probabilities s_1 and s_2 contain, in addition to the source term of micromixing, a correction term which is needed to ensure that the moment closure is predicted correctly. Details on that feature are given in [72] and [88]. The probabilities p_1 and p_2 are transported without additional source terms. The boundary conditions for the simulations performed in the T-CIJM with two inlets and one outlet are chosen consonant with Wang and Fox [88], see Table 3.5. Figure 3.21 helps to get an idea of the initial conditions. The perfect overlap of the two delta functions when s_1 and $s_2 = 0.25$ represents the situation of perfect (micro-) mixing. More information can be found elsewhere [72, 18].

Table 3.5: Boundary conditions used within the DQMOM-IEM two-environment (2E) approach, consonant with Wang and Fox [88].

	Inlet 1	
$p_1 = 1.0$	$\xi_1 = 1.0$	$s_1 = 1.0$
$p_2 = 0.0$	$\xi_2 = 0.0$	$s_2 = 0.0$
	Inlet 2	
$p_1 = 0.0$	$\xi_1 = 1.0$	$s_1 = 0.0$
$p_2 = 1.0$	$\xi_2 = 0.0$	$s_2 = 0.0$

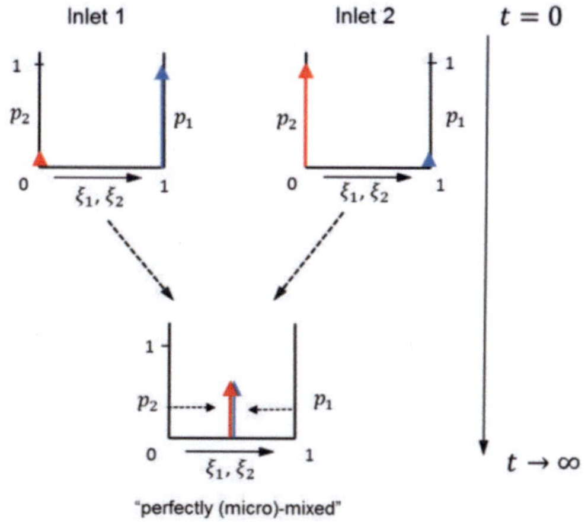

Figure 3.21: Probability density function scheme of the DQMOM-IEM (2E) approach. Initial conditions ($t = 0$) and the case of a "perfectly mixed" system for $t \longrightarrow \infty$ are shown.

The micromixed species concentration for exemplarily chosen components A and B of the initial molar concentration $\tilde{c}_{A,0}$ and $\tilde{c}_{B,0}$ can be estimated by eq. 3.27 and 3.28 using the first moment ($s_1 + s_2$) of the DQMOM-IEM approach.

$$\bar{\tilde{c}}_A = \tilde{c}_{A,0} \cdot (p_1 \xi_1 + p_2 \xi_2) = \tilde{c}_{A,0} \cdot (s_1 + s_2) \quad 3.27$$

$$\bar{\tilde{c}}_B = \tilde{c}_{B,0} \cdot (p_1(1 - \xi_1) + p_2(1 - \xi_2)) = \tilde{c}_{B,0} \cdot (1 - (s_1 + s_2)) \quad 3.28$$

Investigations concerning the influence of a micromixing model implemented into the CFD framework are carried out using the DQMOM-IEM two-environment (2E) model, regarding only the nonreactive mixing case. Thereby, results are compared directly to those achieved with the standard species transport model, which is solved in parallel. If micromixing plays a crucial role, obvious deviations are expected when measuring the mass-weighted and temporally averaged supersaturation in discrete planes along the mixing geometry.

Figure 3.23 shows snapshots of the transport scalars from eq. 3.22 to 3.25. Thereby, the micromixing constant for every grid cell is calculated following eq. 3.21. Resulting concentrations of the ions Ba^{2+} and SO_4^{2-} involved are obtained, as shown in eq. 3.27 and 3.28. These concentrations are then considered when calculating the activity-based supersaturation in every grid cell. Resulting supersaturation values along discrete planes in the mixing zone are shown in Figure 3.24. Thereby, one can see that in both cases regarded, the laminar case and the turbulent case, only infinitesimal deviations between DQMOM-IEM and standard species model can be measured for a simulation period of about $1.5\ s$ of process time.

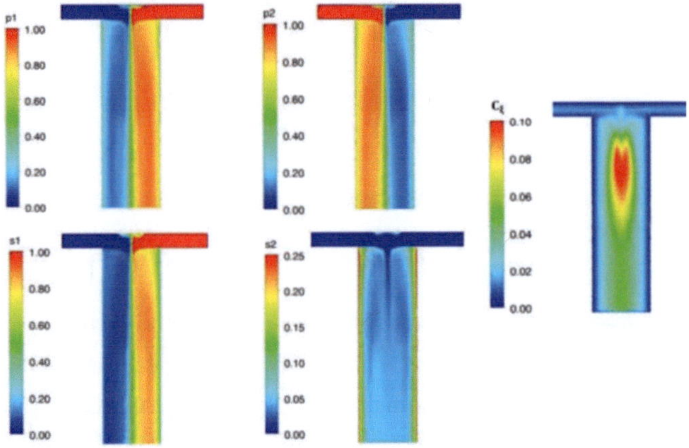

Figure 3.22: Overview of the four scalars transported (eq. 3.22 to 3.25) and the micromixing constant C_ξ calculated in every grid cell at laminar flow conditions in the T-CIJM, $\mathrm{Re_{mix}} = 40$.

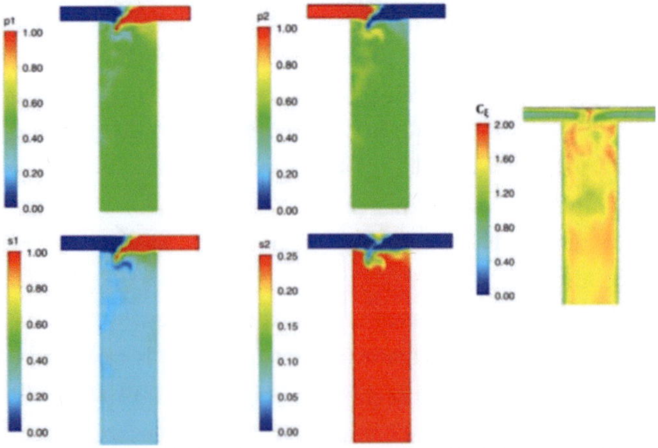

Figure 3.23: Turbulent flow conditions in the T-CIJM, $\mathrm{Re_{mix}} = 4000$.

Small deviations observed in the turbulent case are presumably the result of the highly transient nature of the flow field ($\mathrm{Re_{mix}} = 4000$).

Computational costs increase significantly involving the DQMOM-IEM model, just because the time step necessary for a stable numeric solution has to decrease to values of 10^{-7} to 10^{-9} s. For larger timesteps, the DQMOM-IEM approach delivers, although mathematically correct, unphysical negative values for either one of the weighted probabilities. Despite a bigger effort, no significant influence of the micromixing model on the mixing step could have been shown in the numerical simulation presented.

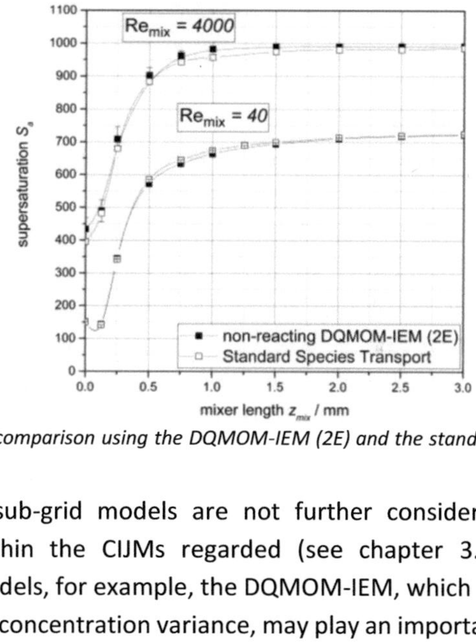

Figure 3.24: Model comparison using the DQMOM-IEM (2E) and the standard species transport model.

In conclusion, sub-grid models are not further considered for the CFD calculations within the CIJMs regarded (see chapter 3.4.2. and 3.4.3). Micromixing models, for example, the DQMOM-IEM, which resolves sub-grid gradients of the concentration variance, may play an important role for larger grid cells.

3.4.2 Coupled computational fluid dynamics - population balance equation (CFD-PBE) approach

As mentioned in chapter 1, supersaturation, as the driving force for primary processes of particle synthesis, such as nucleation and growth (see section

2.3.1 and 2.3.2), is a transient and local quantity. Thereby, flow mechanics and its inherent turbulence are the key element to create initially a mixed environment where such a reaction, followed by a supersaturation buildup, can occur (see section 3.3.1). Hence, particle formation is coupled directly to the flow field and the predominant mixing level of ions involved.

Computational fluid dynamics methods provide a precious way to give predictions about such transient, local mixing states and state variables, such as supersaturation S_a which is a hardly measurable quantity. Moreover, since the solid formation at high supersaturation is a process with very short process times in the range of milliseconds, experimental access is poorly given. One possibility to gain insight into this intensive coupling of fluid dynamics and particle formation is a fully resolved CFD simulation involving all relevant mechanisms.

The starting point of this section is the work of Gavi et al. [66], who modeled the precipitation of $BaSO_4$ with the DQMOM-IEM micromixing closure and a QMOM approach using RANS equations in a stationary mode. The authors could show that this promising efficient QMOM closure for the PBE can handle complex phenomena, such as nonlinear size-dependent growth rates and aggregation, and delivers comparatively good agreement between simulation and experiments. Our work uses the DQMOM approach (see section 2.1.2), which forgoes the inverse error-prone transformation of weights and abscissas into the particular moments in every time step, to solve the PBE, as presented by Marchisio and Fox [16] and Wang and Fox [88]. Moreover, the transient DES-SST-k-ω turbulence model (see chapter 3.2) is used. As experimental investigations in section 3.3.1 could show, simulation of the highly non-steady turbulent behavior in CIJMs, which represents the crucial mixing step, is improved significantly using transient LES [71] or a DES formulation [54]. Therefore, the CFD-PBE approach is implemented into the transient turbulent DES framework to simulate precipitation in a more realistic environment. The activity coefficient model according to Pitzer, see 2.2.3 [23], considering binary and ternary interaction parameters and the formation of complex ions [24, 29], is involved (see section 2.2). Importance is given to the profound implementation of material properties, such as density and viscosity, for the multicomponent mixture (see appendix 8.2.5). Another important issue when simulating the coupled CFD-PBE case is to consider the coupled component balances between solvated reacting ions and the newly built and growing particulate phase. This coupling must be realized with an appropriate source term formulation. Figure 3.19 gives a schematic overview of the source term interplay of liquid and solid phases established for every time step within CFD.

Firstly, the ion fractions which contribute to supersaturation in every grid cell must be scrutinized. Chapter 3.4.1 addresses this question. Thereby, the necessity of a micromixing model that accounts for potentially occurring sub-grid mixing phenomena lowering the effective mixed fraction in a cell is investigated. Since sub-grid models for micromixing for the mesh used play a negligible role in the mixing-dominated $\mathrm{Re_{mix}}$-region ($\mathrm{Re_{mix}} = 40 - 4000$, see Figure 3.24), γ-calculations, supersaturation and component source term coupling is accomplished with the standard species transport.

Once supersaturation is computed, the progress of the weights w_i and abscissa $\mathcal{L}_i$ by the primary rates is calculated within the DQMOM PBE approach. The CFD matching source terms are handed over to these transported scalars of the corresponding solid phases.

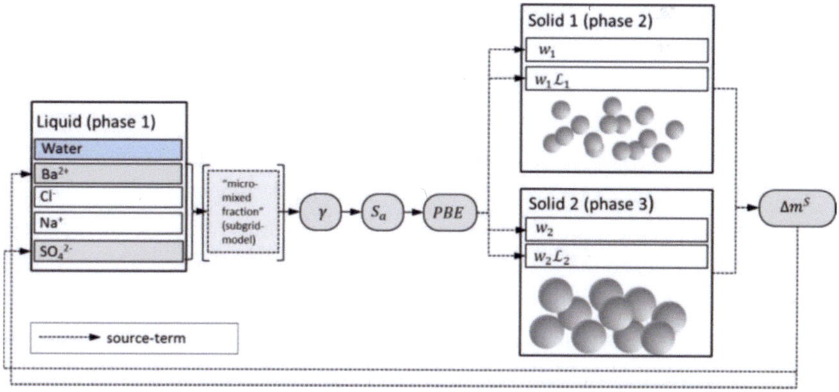

Figure 3.25: Basic scheme of the CFD-PBE coupling, using three Eulerian phases (Liquid, Solid 1 and Solid 2), which is cycled for every numerical time step during transient calculations. Micromixing is neglected (see chapter 3.4.1) and, therefore, is set into paranthesis.

The detailed formulation of the DQMOM procedure can be found in the appendix 8.5.1. Due to the very small time-stepping necessary for a stable solution of the DQMOM algorithm ($10^{-7} - 10^{-9}$ s), multicore calculations are highly computationally cost-intensive (~ days to weeks). An important aspect when setting up the approach properly is the choice of the initial parameters for the PBE-CFD coupling. If the initial value for the particulate phase is chosen to be too large (α_1 or $\alpha_2 \geq 10^{-9}$), the DQMOM algorithm will calculate the wrong particle sizes and a wrong time dependency. This is because, in this case, an artificial shower of nuclei is induced initially into the system that accelerates the ion consumption of the evolving system.

Coupling with the dissolved ionic component fractions is achieved by calculating the mass which is transferred from the liquid to the solid phase Δm^S $(= \Delta \mathcal{M}_3 \cdot \rho^S)$ within the numerical time step and the appropriate reduction of the reacting species concentration. Figure 3.26 and Figure *3.27* show snapshots of the quasi-stationary CFD calculations at $\mathrm{Re_{mix}} = 400$ and 4000 at a barium excess $(\mathrm{R_c} = 5)$. This excess enables the comparison to primary particle sizes obtained experimentally that are not affected by aggregation due to electrostatic stabilization (see chapter 2.4). Quasi-stationarity for the population balance calculation is reached after about two residence times in the mixing zone, which is highly computationally cost-intensive for low Reynolds numbers and a time-stepping necessary of $\sim 10^{-7}$ s. In addition to the two abscissas $\mathcal{L}_1$ and $\mathcal{L}_2$ on a longitudinal cut along the 3D mixing chamber, the volume fraction and the residual supersaturation are shown.

The solution of the PBE for every grid cell in the transient turbulent flow field enables a high local and temporal resolution. In both cases, considering Figure 3.26 and Figure 3.27, the dominant region for particle formation and growth is located on the first two to four millimeters of the mixing zone, where the main turbulence is induced by the impinging jets. Effective supersaturation production is located preferentially straight at the wavy boundary layer of the impinging planes, which is experimentally investigated in section 0 and found as the dominant mixing feature in CIJMs.

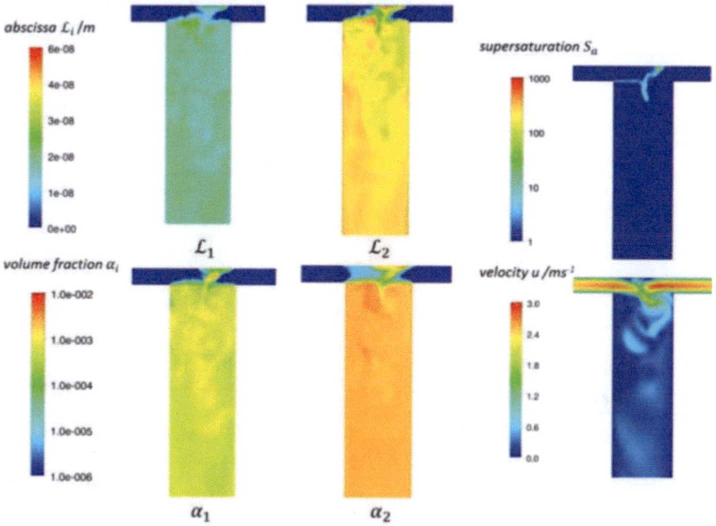

Figure 3.26: Snapshots of the particle abscissas $\mathcal{L}_1$ and $\mathcal{L}_2$, the volume fractions of the crystal phases α_1 and α_2, and the supersaturation and velocity field for $\mathrm{Re}_{\mathrm{mix}} = 400$ in the T-CIJM.

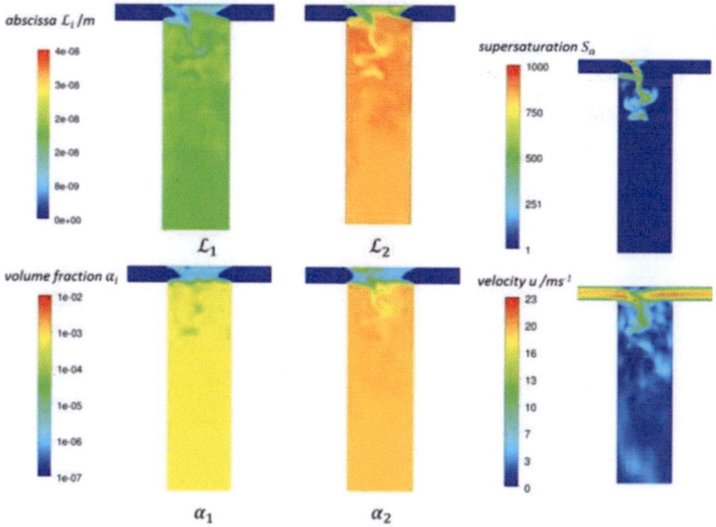

Figure 3.27: Snapshots of the particle abscissas $\mathcal{L}_1$ and $\mathcal{L}_2$, the volume fractions of the crystal phases α_1 and α_2, and the supersaturation and velocity field for $\mathrm{Re}_{\mathrm{mix}} = 4000$ in the T-CIJM.

The data are spatially and temporally averaged (spatially and temporally averaged reduced numeric measurement: STAR NM; details are explained in section 3.4.3) and plotted in Figure 3.28 over the mean residence time to reduce the complexity of these results. Thereby, characteristic particle sizes $L_{1,0}^{P}$ and $L_{3,2}^{P}$ are calculated from the specific moments obtained from the weights and abscissas (see eq. 2.4).

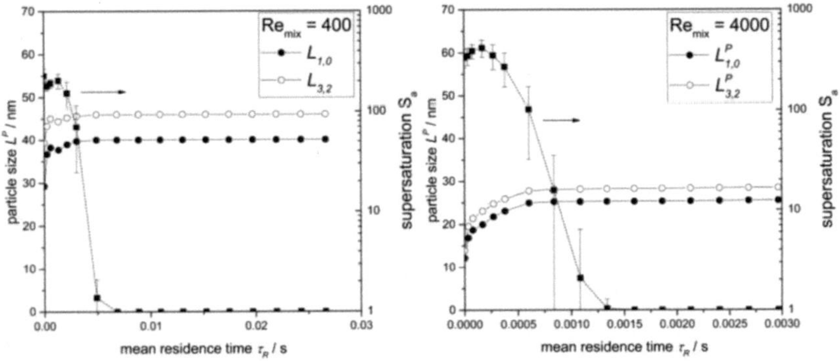

Figure 3.28: Evolution of supersaturation, the characteristic diameters $L_{1,0}^{P}$ and the Sauter diameter $L_{3,2}^{P}$ along the mean residence time in the mixing chamber for $Re_{mix} = 400$ (left) and for $\mathrm{Re}_{\mathrm{mix}} = 4000$ (right).

Figure 3.28 obviously shows that simulated particles for $\mathrm{Re}_{\mathrm{mix}} = 400$ grow bigger than those for $\mathrm{Re}_{\mathrm{mix}} = 4000$. Lower energy dissipation implicates decreasing mixing efficiencies [81, 30]. This immediate lowering impact on the supersaturation buildup (see supersaturation plotted on the secondary axis in Figure 3.29) leads to lower effective supersaturation values at the top of the mixing zone.

Furthermore, a smaller pseudo-stationary supersaturated zone is observed for decreasing $\mathrm{Re}_{\mathrm{mix}}$ numbers as a consequence of the residence time and, thereby, enhanced spatial ion consumption. This behavior can be extracted from Figure 3.29, where snapshots of this supersaturation field during CFD-PBE calculations for different $\mathrm{Re}_{\mathrm{mix}}$ numbers are shown.

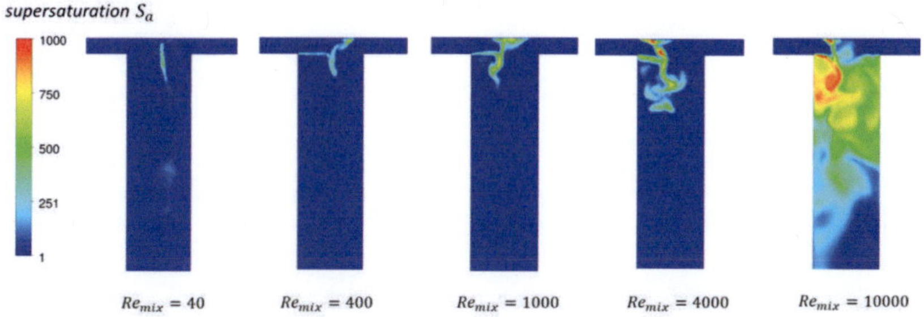

Figure 3.29: Snapshots of the supersaturation field during CFD-PBE calculations in the mixing chamber for different $\mathrm{Re}_{\mathrm{mix}}$ *numbers. From left to right: Increasing flow rate.*

As the driving force is lower, the highly nonlinear nucleation rate (see Figure 2.8) is affected strongly. A change of supersaturation, for instance, of one decade from 1000 to 100 causes a decrease of the homogeneous nucleation rate from values in the scale of 10^{24} to 10^{11} 1/(m^{3}s). For lower flow rates, this significant change in the nucleation rate combined with an increasing residence time in the inhomogeneous zone lead to fewer particles with an increased particle size.

This effect is demonstrated in Figure 3.30, left. With decreasing flow rates, mean particle sizes $L_{50,0}^{P}$, respectively $L_{1,0}^{P}$, extracted at the outlet (which is in equilibrium state) of the T-CIJM increase. Particle sizes fit the mean particle sizes measured experimentally reasonably well. Figure 5.27, right, demonstrates exemplarily the node representation received from CFD compared to the discrete distribution measured experimentally.

Two more interesting effects are observed considering Figure 3.29 and Figure 3.30. Supersaturation exists all along the interface in the mixing chamber for the strictly laminar regime where the two reactant streams only weakly interact. This long-lasting segregated state of the two fluid streams may cause the dramatic changes in the particle sizes as well as the broad dispersion of products observed experimentally (see Figure 3.10). The CFD-PBE method presented gives important evidence on this behavior, but does not fully picture the experiments, since, in reality, the aggregation processes occurring are not considered (see also later in section 3.4.4.1).

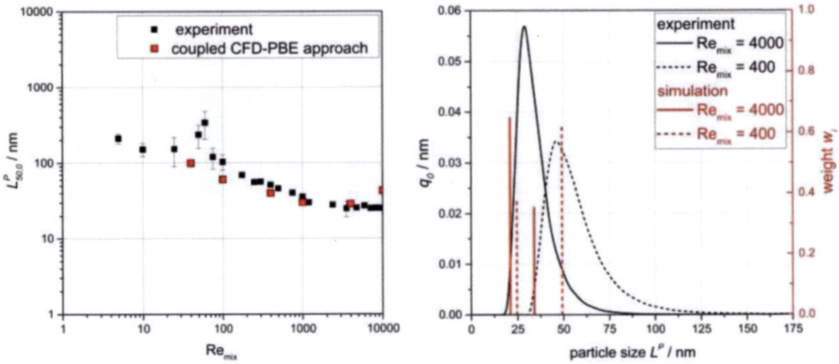

Figure 3.30: Left: Overview of experimental results and simulated mean particle diameters $L_{50,0}$ in the T-CIJM at different flow rates; right: Nodes representation gained at the mixer outlet for $\mathrm{Re}_{\mathrm{mix}} = 400$ and 4000 compared to experiments.

An increasing particle size within CFD is "measured" for high $\mathrm{Re}_{\mathrm{mix}}$ numbers ($\mathrm{Re}_{\mathrm{mix}} = 10\ 000$) due to the very high impulse into the mixing chamber (see Figure 3.29). This effect is not proven experimentally. However, it should be pointed out that no cavitation effects are considered in CFD; the effects, in reality, insert an additional important contribution to mixing (existent for $\mathrm{Re}_{\mathrm{mix}} > 4000$, see section 3.3.1.4).

Conclusions

This chapter elaborates on a coupled 3D CFD-PBE approach for predicting the influence of mixing on the resulting PSD for the model system $BaSO_4$. Thereby, a momentum-based description (DQMOM) for the particulate phase is chosen to reduce the computational effort. It can be well demonstrated how the mixing conditions affect the supersaturation field as the driving force and, in conclusion, the mean particle sizes. The method gives a very detailed insight into the coupling of fluid mechanics and particle formation. However, computational load for such simulations is high. Multicore simulations are in the range of days to weeks due to the very small time stepping necessary and are, therefore, not feasible for fast flow sheet modeling aimed at, see chapter 1.

3.4.3 STAR NM – spatially and temporally averaged reduced numeric measurement

The CFD may be an important key instrument to gain access to process-dominating phenomena. However, as demonstrated in chapter 3.4.2, the spatial and temporal resolution of the simulation must be set adequately high. This expands the computational effort enormously. Because of this high level of detail, there is no possibility to derive reduced, fast short-cut models which span over a wide parameter space.

The new approach presented in this chapter, which is called STAR NM, tackles this issue. It is demonstrated how coarse-graining of such a sensitive and very fast mixing process can be done in the case of precipitation.

The STAR NM method is based on preliminary studies of Li et al. [91]. Li et al. [92] demonstrated how this coarse-graining method (scale separation) can be adapted to batch processes by extracting spatially and temporally averaged reduced, numerically "measured" growth rates within a spray granulation process. Integral rates (e.g. growth rates) are measured for the granulation chamber and are exported from CFD to a CFD-external fast 1D population balance solver, which quickly the distribution solves only considering the internal coordinate particle size (see eq. 2.1) for a rather long period of process time (several seconds or even tens of minutes). Thereafter, new process states (e.g. the size distribution and mass of solids) are patched back to CFD to update the rates "measured" numerically. With this alternating approach, switching from CFD to PBE and back again, they could simulate process times of several minutes to hours with a high accordance to experimental data. Fully coupled CFD-PBE simulations, as presented in section 3.4.1, are not able to stretch out for such long process times.

In this section, how to transfer this STAR NM method to problems that must be solved by differential balances instead of the integral ones considered by Li et al. [91] is demonstrated. The measurement of nucleation and growth rates for precipitation does not make sense due to the direct coupling and back-coupling with the component balances of reacting ions. In the following, it is demonstrated how this fruitful interplay between CFD and

PB, and for differentially balanced processes can be created. This is achieved by measuring only the driving force without reaction (or the mixed ion concentration that actually contributes to this driving force) and the residence time distribution in the apparatus regarded. Since the expression "measured" is commonly dedicated to experimental investigations, CFD "measurements", which depend on the accuracy of simulations, are indicated with quotation marks. The general procedure of the STAR-NM approach for precipitation can be summarized in the following three steps:

1. *STAR numeric "measurement" of either the mixed components that contribute to the supersaturation or direct STAR numeric "measurement" of the supersaturation, which is only reasonable if the free lattice ion ratio is assumed to be almost constant.*
2. *STAR numeric "measurement" of the residence time distribution in the apparatus.*
3. *Export of the data which are achieved in step 1 and 2 to a PBE-solver. Solving of the population balance by the PBE-solver including the data measured in step 1 and 2.*

In the following section, a detailed description will be given of this procedure (points 1 to 3) for the case of precipitation of $BaSO_4$ in a T-mixer. The method may also work for other reaction engineering problems, such as fast parallel reactions. Important results and the potential of the approach will also be discussed.

3.4.3.1 Step 1: "Measurement" of the mixed ions or supersaturation buildup

The basic idea of the new coarse-graining approach is the "measurement" of experimentally inaccessible state variables, such as the "mixed" component fractions or the local supersaturation, by using transient CFD simulations in combination with an appropriate thermodynamic model. This data set is then averaged with respect to time and lateral space.

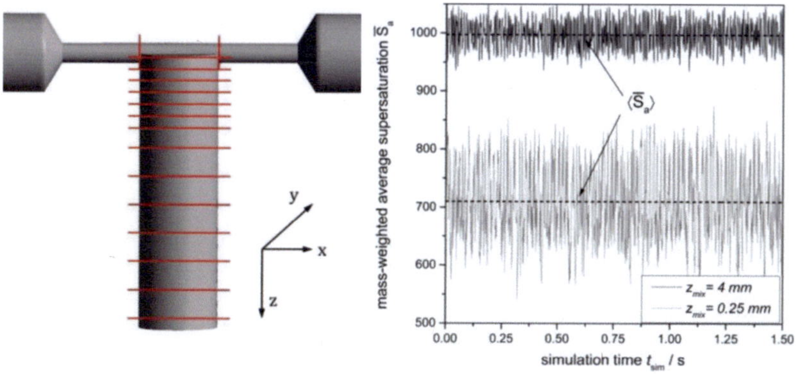

Figure 3.31: Left: Discrete "measurement" planes chosen wherein the topmost plane is defined as $z_{\mathrm{mix}} = 0$. Transient quantities are, on the one hand, measured mass-weighted averaged and, on the other hand, averaged in time, right: Measured mass-weighted averaged supersaturation values $\bar{S}_a$ in two discrete planes ($z_{\mathrm{mix}} = 0.25\ mm$ and $z_{\mathrm{mix}} = 4\ mm$) in the mixing chamber for $S_{a,\mathrm{nom}} = 1000$, $\mathrm{R_c} = 5$ and $\mathrm{Re_{mix}} = 4000$. The red lines indicate the spatially and temporally reduced value $\langle\overline{S_a}\rangle$.

Hence, the supersaturation level here is "measured" over several seconds of simulated process time in discrete planes along the mixing geometry (see Figure 3.31). A final temporal averaging over at least 1 s of process time of the highly transient plane values leads to a single set of data for the specific process conditions chosen ($\mathrm{Re_{mix}}, S_{a,\mathrm{nom}}, \mathrm{R_c}, \mathrm{M_R}$) (see Figure 3.32). The momentum ratio $\mathrm{M_R}$ of the reactant streams entering the mixing chamber is set to a value of one and is not varied in this work. Mathematical functions that correlate the supersaturation buildup "measurements" in the CIJM (Figure 3.32) read as

$$\langle\bar{S}_a\rangle(S_{a,\mathrm{nom}}, z_{\mathrm{mix}}) = A_1 \cdot e^{\left(-\frac{z_{\mathrm{mix}}}{B_1}\right)} + S_{a,\mathrm{nom}} \qquad 3.29,$$

wherein the fit parameters A_1 and B_1 are $f(\mathrm{Re_{mix}}, S_{a,\mathrm{nom}}, \mathrm{R_c})$, z_{mix} is the mixing length and $S_{a,\mathrm{nom}}$ is the value adjusted via the reactant concentration and which is reached in the case of perfect mixing.

Results shown in this section are exemplarily selected ones to illustrate the general procedure of STAR NM. Basically, data measured in the CIJM are dependent on the four parameters $S_{a,\mathrm{nom}}, \mathrm{R_c}, \mathrm{M_R}$ and $\mathrm{Re_{mix}}$ (respectively $\bar{\varepsilon}$)

mentioned. Free lattice ion ratios $\mathrm{R_c} < 1$ (see eq. 2.28), for instance, as well as smaller $\mathrm{Re_{mix}}$ numbers lead to a slightly faster exponential growth. These influences of the free lattice ion ratios $\mathrm{R_c}$ are the result of the strong nonlinear dependency of the activity coefficient for the nonstoichiometric case ([29], see also Figure 3.25). Lower $\mathrm{Re_{mix}}$ numbers and, thus, a lower entrance impulse, but longer residence time leads to a slightly shorter mixing zone. Correlation parameters in the laminar and the transition regime must be adapted to decreasing mixing intensities and, thus, longer mixing times. Interested readers may extract such influencing effects from measurements presented in the appendix (8.2.3). Figure 3.32 shows the "measured" dependency of the parameter $\mathrm{S_{a,nom}}$ on the mixing kinetics.

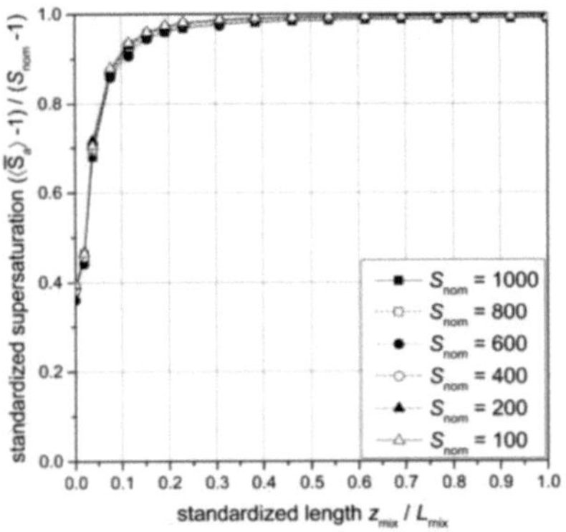

Table 3.6: Correlation coefficients to be used in eq. 3.29

(Figure 3.32)	*for* $\mathrm{Re_{mix}} = 1000, \mathrm{R_c} = 1, \mathrm{M_R} = 1$
A_1	$-0.6659 \cdot S_{a,\mathrm{nom}}$
B_1	0.3494

Figure 3.32: Dimensionless "measured" master curve for the supersaturation buildup at different nominal supersaturation levels $S_{a,\mathrm{nom}}$ at constant $\mathrm{Re_{mix}} = 1000$, $\mathrm{R_c} = 1$ and $\mathrm{M_R} = 1$ measured within CFD with $L_{\mathrm{mix}} = 7\ mm$.

Thereby, this dimensionless master curve presented ($\mathrm{Re_{mix}} = 1000$, $\mathrm{R_c} = 1$, $\mathrm{M_R} = 1$) is representative of the turbulent supersaturation buildup along the mixing length z_{mix} for every $S_{a,\mathrm{nom}}$, and can be correlated by eq. 3.29. The correlation parameters are shown in *Table 3.6*. Experimental evidence that strongly support the validity of these measurements are given in section 3.3.1.3. Both measurements executed, on the one hand, the level of supersaturation, and, on the other hand, the intensity of segregation used as a mixing indicator, deliver congruent results.

3.4.3.2 Step 2: "Measurement" of the residence time distribution

The residence time behavior of the CIJM investigated is extracted by tracking about 2500 massless Lagrangian particles during their flow through the mixing volume (Figure 3.33).

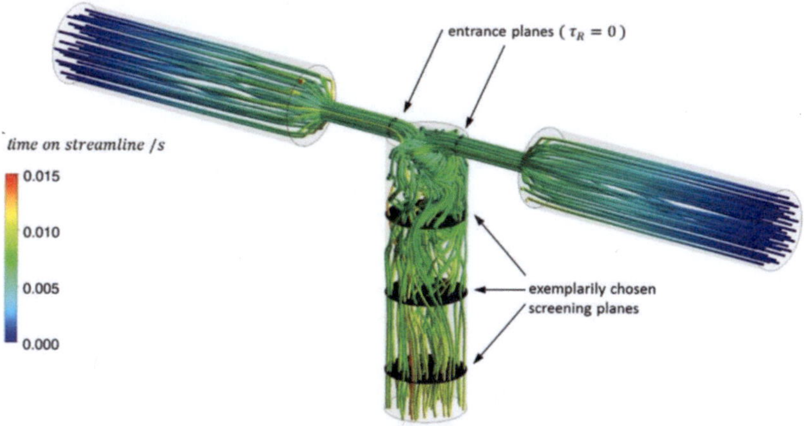

Figure 3.33: Residence time measurement in the T-mixer for $\mathrm{Re}_{\mathrm{mix}} = 4000$ *from the inlet into the mixing chamber to discrete planes along the flow direction*

The time of flight required for each particle is measured from the entrance of the mixing chamber to specific screening planes (see Figure 3.33) in the mixing chamber, as indicated. These planes are specified in Figure 3.31, left. Thereby, a mean cumulative residence time distribution for every plane in the downstream part of the mixer is obtained (see Figure 3.34, left).

These measurements are reduced once again in the framework of STAR NM to their median value τ_{50} and two further nodes τ_{10} and τ_{90} (see the three arrows in Figure 3.34, left). Consequently, the mixing length can be plotted versus the median value of the residence time τ_{50} for the inert tracer particles that are tracked while traveling along the mixing geometry (see Figure 3.34 and Figure 3.35).

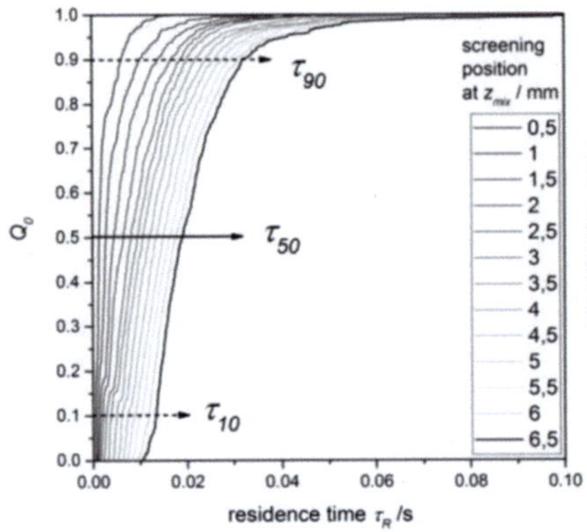

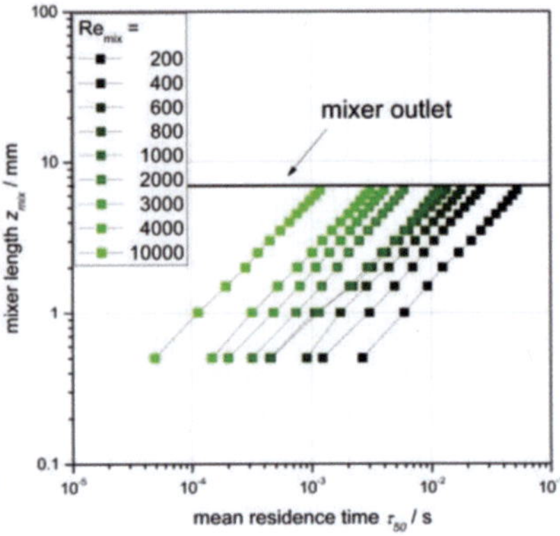

Figure 3.34: Left: Example of the cumulative residence time distribution from the entrance planes ($\tau_R = 0$) to the specific screening plane (see Figure 3.33, right) when inserting 2500 tracer particles at $\mathrm{Re_{mix}} = 400$; right: Mixing length vs. mean residence time at different $\mathrm{Re_{mix}}$ numbers for the CIJM.

Data in certain flow regimes can be fitted well with a power-law function. The fit parameters a_2 and b are $f(\mathrm{Re_{mix}})$.

$$z_{\mathrm{mix}} = a_2 \cdot \tau_{50}^{b} \qquad 3.30$$

Selected values for the correlation parameters a and b in eq. 3.30 are given in Table 3.7. The correlation parameters given describe very well the mean residence time behavior of the T-mixer investigated in the turbulent regime for $\mathrm{Re_{mix}} > 1000$. A detailed overview of the correlation coefficient in the laminar and transition regime is given in the appendix.

Table 3.7: Correlation coefficients for the mean turbulent residence time τ_{50}, to be used in eq. 3.30

Correlation coefficient	$\mathrm{Re_{mix}} > 100$
a_2	$23.4843 \cdot \mathrm{Re_{mix}^{0.44579}}$
b	0.86965

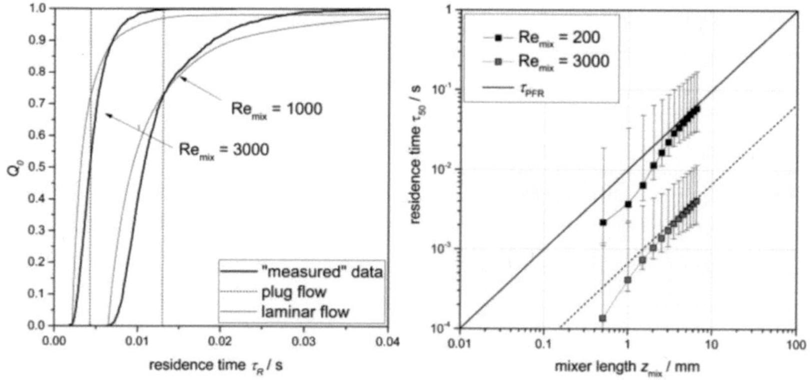

Figure 3.35: Left: Exemplarily chosen residence time distributions for $\mathrm{Re}_{\mathrm{mix}} = 1000$ *and* 3000 *at* $z_{\mathrm{mix}} = 6.5\ mm$ *(mixer outlet) compared with theoretically predicted residence time distributions for an ideal plug flow and for laminar pipe flow (Hagen-Poiseuille); right: Comparison of the mean residence time measured to a pipe plug flow for two exemplarily chosen* $\mathrm{Re}_{\mathrm{mix}}$ *numbers. The error bars mark the dispersion of the flow in terms of* τ_{10} *and* τ_{90}*.*

Figure 3.35, left, shows two exemplarily chosen residence time distributions for $\mathrm{Re}_{\mathrm{mix}} = 1000$ and 3000 at the mixer outlet which are for comparison supplemented by two limiting cases. On the one hand, the typical plug flow behavior; on the other hand, the laminar flow distribution resulting from the Hagen-Poiseuille flow profile is plotted. As can be seen, the real apparatus characteristics lie somewhere in between. The CFD results show that the measured τ_{50} deviates from this idealized plug flow assumption even on the first $1 - 4\ mm$ and runs lower than τ_{PFR} until it converges to (see Figure 3.35, right). Through the high entrance impulse of the fluids in the mixing chamber, most particles pass this distance in a velocity faster than $\bar{u}$. The first millimeters in the mixing chamber are highly turbulent (Figure 3.33) and lead to a wide particle distribution due to back-mixing of some streamlines. In the mixing progress, the mean residence time converges to τ_{PFR}. Furthermore, Figure 3.35, right, shows that deviations of τ_{50} from the idealized plug flow reduce for higher $\mathrm{Re}_{\mathrm{mix}}$. An evaluation of the Bodenstein number Bo to quantify the relationship of convection and axial dispersion supports this circumstance (see appendix).

3.4.3.3 Step 3: Combining computational fluid dynamics "measurements" and population balance

One aspect of complexity when solving the PBE (see eq. 2.1) originates from coupled component balances for the species i involved ($i = Ba^{2+}, SO_4^{2-}$) that must be solved for each numerical time step Δt. The coupling of reacting ions that change phases from liquid to solid can be described with eq. 3.31.

$$\tilde{c}_i(t+\Delta t) = \tilde{c}_i(t)$$

$$-\tilde{\rho}^S \cdot \left(\sum_{k=1}^{n} \left(\pi (\bar{L}^{P_k})^2 \cdot \frac{G(\bar{L}^{P_k}, S_a)}{2} \cdot n_k \right) + \left(\frac{\pi}{6} (L^P_{\text{crit}}(S_a))^3 \cdot B_{\text{hom}}(S_a) \right) \right) \cdot \Delta t \qquad 3.31$$

The index k denominates the specific size class of the particle distribution, which is discretized into n classes.

For the ideally mixed case at $t = 0$, $\tilde{c}_i(t = 0) = \tilde{c}_{i,\text{nom}}$. Calculations terminate when either one of the solvated ionic reactants $\tilde{c}_i$ is fully consumed.

The effect of nonideal mixing is implemented using the data correlated from the STAR NM approach combining eq. 3.29 and 3.30. Thereby, the "measured" value $\tilde{c}_{i,\text{STAR NM}}$ in the first screening plane in the mixer at $z_{\text{mix}} = 0$ represents the starting point ($t = 0$) of the PB calculations.

The method assumes perfect advection of the nanocrystals ($\text{St} \ll 1$) [51]. This implies that the particle relaxation time is small and the particles follow the flow. Thereby, the evolution of $S_a = f(t)$ – assuming only mixing – for a set of fixed parameters $S_{a,\text{nom}}$, R_c, and Re_{mix}, is calculated along the process time t ($\triangleq$ e.g. τ_{50}). Inverse calculations from the supersaturation and R_c present at t leads to the ions involved $\left(\tilde{c}_{i,\text{STAR NM}}(t)\right)$. It is simplified for the CIJMs, assuming that R_c relaxes very fast on its adjusted value and is taken as constant (e.g. $\text{R}_\text{c} = 5$). This assumption works out well for CIJMs. For other apparatuses, such as the ST discussed in section 5.5 with strongly asymmetric stoichiometric conditions, this procedure must be handled carefully. Thus, it is better to correlate the mixed concentrations $\tilde{c}_{i,\text{STAR NM}}$ directly independent of the residence time.

The very important step is the consideration of the superposing effects of supersaturation buildup and the depletion of supersaturation due to particle formation. This means the prevailing supersaturation has to be calculated by estimating a "mixed" $\tilde{c}_i(t)$ in eq. 3.31, taking into account the supersaturation buildup predicted by the mixing correlation (combining step 1 and 2 of the STAR NM approach) that is decreased by the ions consumed in earlier time steps.

$$\tilde{c}_i(t) = \tilde{c}_{i,\mathrm{STAR\,NM}}(t) - \int_0^t \frac{\partial \tilde{c}_{i,\mathrm{depl}}}{\partial t}(t)\, dt \qquad 3.32$$

Together with the unchanged inertial ion concentrations that contribute to the ionic strength I (see *Table 2.2*), the updated, effectively available concentrations of the two reacting ions Ba^{2+} and SO_4^{2-} (eq. 3.32) are reinserted into the activity coefficient model. A new supersaturation is calculated.

- **ideally mixed case:** $\tilde{c}_i(t = 0) = \tilde{c}_{i,\mathrm{nom}}$.
- **nonideally mixed case:** $\tilde{c}_i(t = 0) = \tilde{c}_{i,\mathrm{STAR\,NM}}(t = 0)$

Figure 3.36, left, shows the progression of the saturation level along the CIJM for the nominal supersaturation $S_{a,\mathrm{nom}} = 1000$ and considers the depletion of supersaturation due to the particle formation process. Idealized population balance approaches (dotted line) assume instantaneous mixing at the feed entrance point [93]. Data generated involving the supersaturation evolution "measured" with the STAR NM show that this idealization does not hold. Contrary to the idealized calculation, this model shows that supersaturation buildup takes place even during the particle formation. Thus, the nominal supersaturation level of $S_{a,\mathrm{nom}} = 1000$ is never reached.

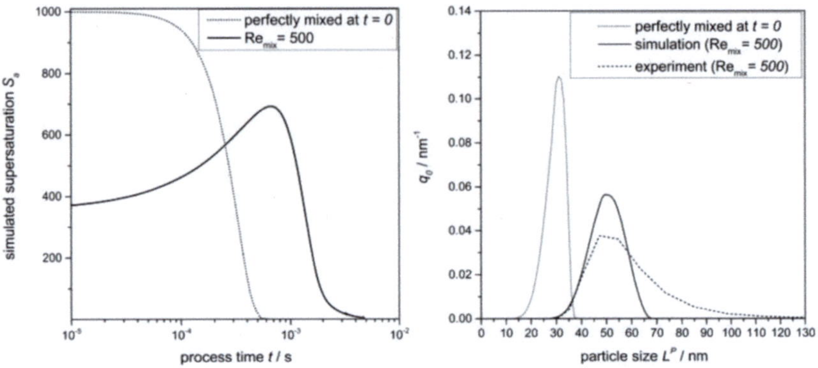

Figure 3.36: Left: Calculated time-dependent supersaturation during particle formation process for $S_{a,\mathrm{nom}} = 1000$*; Right: Simulated PSDs for* $S_{a,\mathrm{nom}} = 1000$*,* $\mathrm{R_c} = 5$ *and* $\mathrm{Re_{mix}} = 500$ *compared to the appropriate experimental size distribution.*

Lower effective supersaturation levels lead to a lower nucleation rate and, therefore, to bigger particles, since all mass is distributed onto fewer crystals. Figure 3.36, right, shows the resulting PSD, which has a high degree of compliance to experiments. Because the averaged mean values are used both at supersaturation buildup and residence time measurement (τ_{50}), the width of the distribution observed experimentally cannot be described properly.

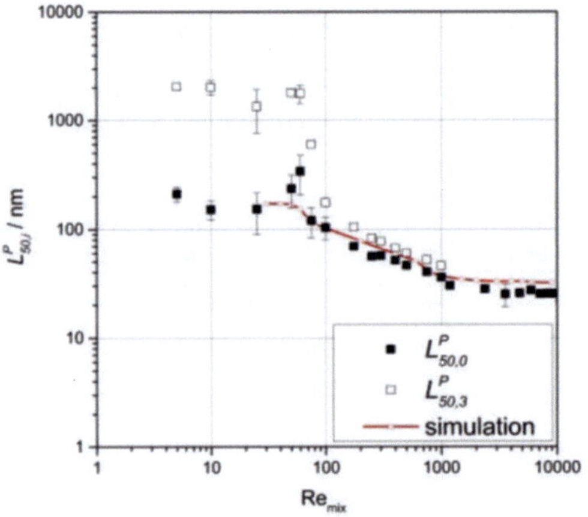

Figure 3.37: Influence of the $\mathrm{Re_{mix}}$ *number in the T-mixer, investigated on the mean particle size* $L^P_{50,0}$ *of the product distribution. The experimentally measured* $L^P_{50,3}$ *is added for comparison, which is not predicted well for the laminar case (*$S_{a,\mathrm{nom}} = 1000$ *and* $\mathrm{R_c} = 5$*).*

Figure 3.37 shows the mean volumetric particle size $L_{50,0}$ of the crystals precipitated. Thereby, the PSD shows a pronounced dependence on the Re number. Simulated mean particle sizes calculated with the STAR NM described above are in good agreement with the experiments. The next two subsections address the two open questions regarding Figure 3.36 and Figure *3.37*. On the one hand, the tailing of the PSDs (Figure 3.36, right) is discussed, using extended residence time information from STAR NM step 2. On the other hand, the cause of strongly deviating mean particle sizes $L^P_{50,0}$ and $L^P_{50,3}$ in the laminar regime for $\mathrm{Re_{mix}} < 100$ is investigated.

3.4.3.4 Influence of the residence time dispersion

Figure 3.36, right, illustrates that the larger particles of the experimental PSD cannot be predicted correctly when only taking the τ_{50} value of the residence time spectrum into account. However, the CFD particle tracking provides the entire residence time spectrum. Figure 3.38, left, shows the residence time distribution for $\mathrm{Re_{mix}} = 500$. Moreover, mean particle sizes are calculated (STAR NM, step 3) considering τ_{10}, τ_{50} and τ_{90} as the dominant time for the supersaturation buildup (see squares in Figure 3.38).

The first-order interpolation between this characteristic particle size nodes at τ_{10}, τ_{50} and τ_{90}, and the superimposition with the residence time spectra lead to Figure 3.38, right.

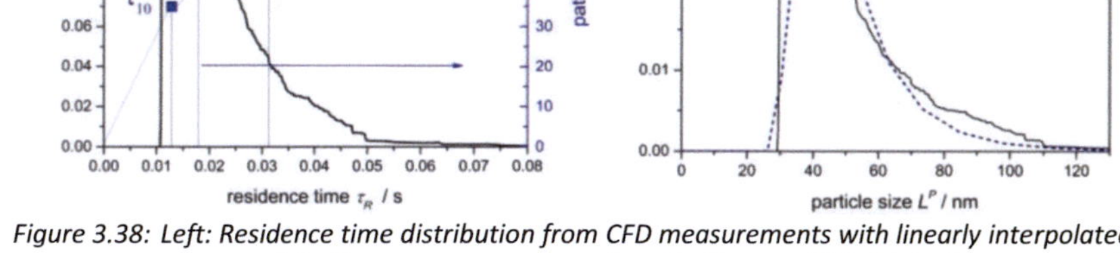

Figure 3.38: Left: Residence time distribution from CFD measurements with linearly interpolated mean particle diameters at τ_{10}, τ_{50} and τ_{90} for $\mathrm{Re_{mix}} = 500$. Right: Discrete particle distribution, experiment and simulation for $S_{a,\mathrm{nom}} = 1000$, $\mathrm{R_c} = 5$ at $\mathrm{Re_{mix}} = 500$.

In other words, every particle size is weighted with its probability to be formed during the process. Thus, it can be shown that the broadening of the experimental PSD in the transition and turbulent regime is solely related to the characteristic residence time spectrum of the precipitation apparatus.

3.4.3.5 Laminar mixing in CIJMs

The mean, number-weighted particle diameter $L_{50,0}$ and the mean, volume-weighted particle diameter $L_{50,3}$ in Figure 3.37 are similar in the transition and turbulent regime and differ significantly in the laminar regime.

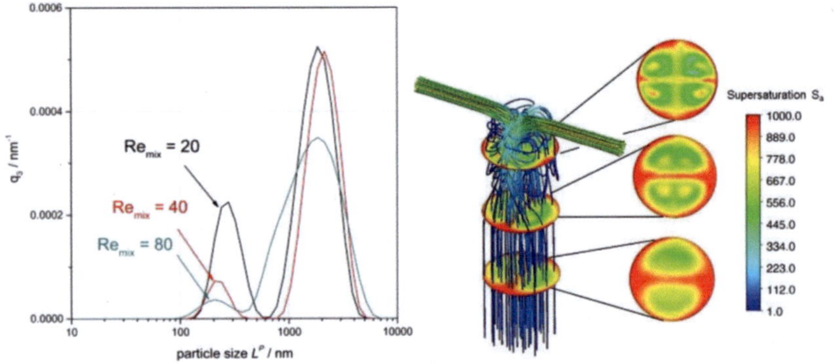

Figure 3.39: Left: PSDs measured in the laminar regime of the CIJM for three different $\mathrm{Re}_{\mathrm{mix}}$ *numbers. Right: Supersaturation field without particle formation and streamlines in the laminar case at* $\mathrm{Re}_{\mathrm{mix}} = 40$ *extracted from CFD.*

$\mathrm{Re}_{\mathrm{mix}} \geq 100$ is detected experimentally and simulative as the onset of fluctuations of the impinging point (see section 3.3.1.4, transition to engulfment flow). In this case, where fluctuations lead to an enhanced mixing intensity, a monomodal distribution is measured. Bimodal and higher multimodal distributions are observed in experiments in the strictly laminar regime, where fluid streams do not exchange matter intensively and the mixing intensity is low ($\varepsilon < 10\ W/kg$), see Figure 3.39.

Figure 3.40: SEM images of laminar-precipitated particles in the T-CIJM investigated; left: $\mathrm{Re}_{\mathrm{mix}} = 20$*, right:* $\mathrm{Re}_{\mathrm{mix}} = 80$.

The SEM images in Figure 3.40 strongly support the light scattering results shown in Figure 3.39, left. As can be seen, at least two fractions of particles can be observed. A fine fraction below $200\ nm$ and some significantly bigger particles that are more numerous for the strictly laminar case ($\mathrm{Re}_{\mathrm{mix}} = 20$).

Depending on the definition of the mean diameter, the one or the other peak is weighted more strongly and describes the behavior observed more closely. The CFD simulation helps to get an insight into such a laminar mixing situation. Figure 3.39, right, shows the supersaturation field in the mixer without reaction (STAR NM, step 1) at $\mathrm{Re}_{\mathrm{mix}} = 50$ and the calculated fluid trajectories newly built crystals would follow in the mixing zone.

In addition to a weak mixing period in the mixer-head, the streamlines are aligned strictly parallel. This shows that particle trajectories undergo completely different supersaturation fields than in the turbulent regime. As the segregated supersaturation fields in the cross-sections of the mixing chamber show, where distinct regions of high and low supersaturation values exist (see Figure 3.39, right), at least a bimodal distribution is to be expected. One effect that can be even more dramatic is the fact that one half of the mixer in the laminar case is filled with a mixture containing a sulfate excess, whereas the other half is dominated by an excess of barium. The barium ions tend to stabilize the primary particles electrostatically, as illustrated in Figure 2.10 [21]. Particles in regions of sulfate excess may aggregate and form significantly bigger associates (see section 2.4). These effects are not considered in the simulations.

3.4.3.6 Time scale analysis using STAR NM

Spatially and temporally averaged CFD "measurements" of characteristic attributes, such as the supersaturation buildup in CIJMs, give access to time dependencies that are not available experimentally. Thus, one can define a characteristic mixing length and a characteristic mixing time τ_{mix}. It has been decided to use a tangent construction (Figure 3.41, left) in the steepest point of the supersaturation buildup measured (STAR NM, step 1) for all flow rates investigated to evaluate this characteristic time. This procedure combined with the residence time information (STAR NM, step 2) leads to Figure 3.41, right.

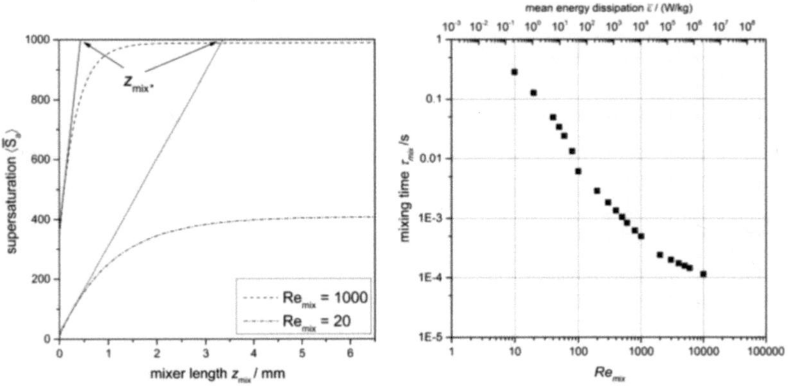

Figure 3.41: Extracted characteristic mixing times from the supersaturation buildup "measurement" with the STAR NM approach; left: Tangent construction to find the characteristic mixing length; right: Overview of all data points extracted.

The same procedure is applied to the depletion curve of supersaturation (see Figure 3.42).

Constructing a tangent in the inflection point of these curves, τ_{depl} can be defined as a characteristic time for the primary processes of particle formation for the model system $BaSO_4$. The temporal evolution of supersaturation is calculated with the proposed population balance model assuming nucleation, growth and perfect mixing at $t = 0$ (see also Figure 3.36).

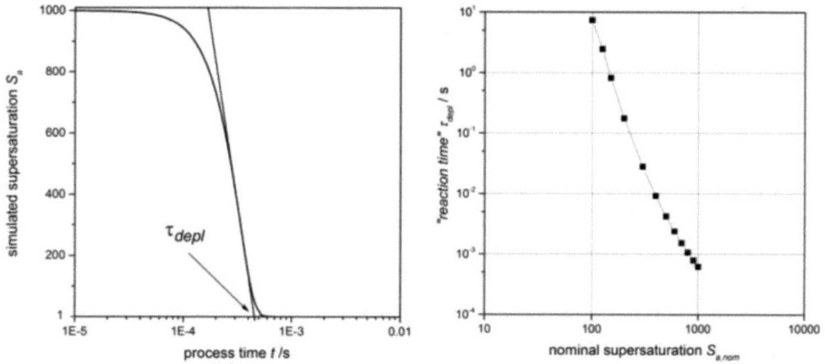

Figure 3.42: Estimation of the characteristic "reaction times" τ_{depl} from simulated supersaturation depletion curves of $BaSO_4$ assuming perfect mixing at $t = 0$. Left: Tangent construction for the evaluation of τ_{depl}; Right: Dependency of τ_{depl} on the nominal supersaturation.

Thus, the PBE approach is used to calculate a hypothetical time for solids formation [5]. The values of τ_{depl} are certainly dependent on the reacting system investigated, whereas τ_{mix} is a function of the apparatus and the process parameters chosen. Figure 3.42 shows the strong dependence of τ_{depl} on the nominal supersaturation for the system $BaSO_4$.

$$\mathrm{Da_t} = \frac{\tau_{\mathrm{mix}}}{\tau_{\mathrm{depl}}} \quad 3.33$$

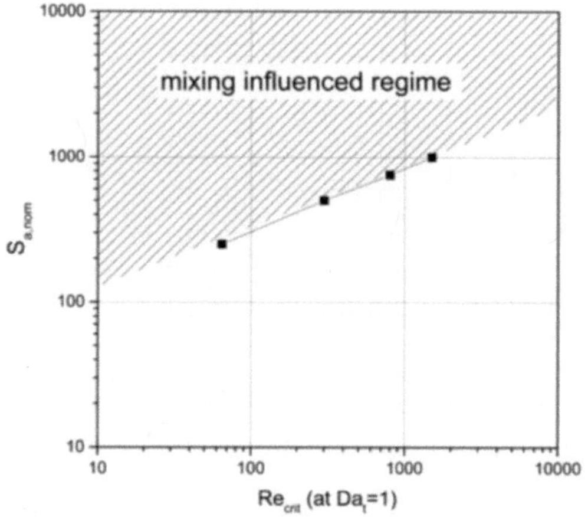

Figure 3.43: Nominal supersaturation vs. critical Re *number for* $\mathrm{Da_t} = 1$.

The characteristic time scale of both the competing steps, mixing and particle formation, can be used to estimate the turbulent Damkoehler number $\mathrm{Da_t}$ which is defined consonant with eq. 3.33.

A critical $\mathrm{Re_{mix,crit}}$ number for the CIJM can be calculated for a given supersaturation $S_{a,\mathrm{nom}}$ of the model system $BaSO_4$ where $\mathrm{Da_t} \overset{!}{=} 1$ (see Figure 3.43). Mixing and particle formation are equally fast at this threshold value. Selecting, for instance, a supersaturation value of $S_{a,\mathrm{nom}} = 1000$, then $\mathrm{Re_{mix,crit}} = 1600$, which relates well to the critical flow rate above which mixing no longer significantly influences the precipitation process, see Figure 3.37. Hence, time scales extracted from the STAR NM approach enable a prospective classification of precipitation processes into a regime which is masked by mixing phenomena ($\mathrm{Da_t} > 1$) and a regime which is not affected by mixing ($\mathrm{Da_t} < 1$).

3.4.3.7 Energy dissipation rates in CIJMs

One important aspect that is required for a predictive, mixing theory-based model is the evaluation of the mean energy dissipation rate that dominates the mixing step in the mixer chamber. This value directly affects meso- and micromixing rates (see also subsequently in chapter 3.4.4). The topic on how to calculate the correct energy dissipation has been discussed manifoldly in the last decade [52, 55]. The fundamental idea all contributions involve is that the energy dissipation at the impinging point is generated by the entrance impulses of the jets into the mixing chamber. Sultan et al. [59] could show that a high entrance impulse, realized with small jet diameters (d_{mix} / $d_{\mathrm{jet}} \geq 4$), improves the energy dissipation rate ($\varepsilon > 10^4\ W/kg$) and the vortex formation that supports mixing in the impinging zone. We focus on two approaches literature provides to calculate the energy dissipation. All data are compared to mass-weighted and temporally averaged energy dissipation values from STAR NM.

Johnson and Prud'homme [52] assume that ε is a rate of energy P inserted into the mixer on a mass in the volume V_{mix} over which the energy is dissipated. Their approach, based on this energy input, reads as eq. 3.34, wherein the power input is calculated with the kinetic energy of the entering jet streams 1 and 2 (method 1).

$$\bar{\varepsilon}_{\text{method1}} = \frac{P}{\rho_{\text{mix}}^{L} \cdot V_{\text{mix}}} = \frac{\left(\frac{1}{2}\dot{M}_{\text{jet1}}\bar{u}_{\text{jet1}}{}^{2} + \frac{1}{2}\dot{M}_{\text{jet2}}\bar{u}_{\text{jet2}}{}^{2}\right)}{\rho_{\text{mix}}^{L} \cdot V_{\text{mix}}} \quad 3.34$$

Siddiqui et al. [55] refined this approach and balanced the overall mechanical energy also involving the kinetic energy of the outgoing stream as well as pressure drops in the mixing chamber Δp (eq. 3.35 and 3.36) (method 2).

$$\bar{\varepsilon}_{\text{method2}} = \frac{(\dot{V}_{\text{jet1}} + \dot{V}_{\text{jet2}})\Delta p + \left(\frac{1}{2}\dot{M}_{\text{jet1}}\bar{u}_{\text{jet1}}{}^{2} + \frac{1}{2}\dot{M}_{\text{jet2}}\bar{u}_{\text{jet2}}{}^{2} - \frac{1}{2}\dot{M}_{\text{mix}}\bar{u}_{\text{mix}}{}^{2}\right)}{\rho_{\text{mix}}^{L} \cdot V_{\text{mix}}} \quad 3.35$$

$$\Delta p = \frac{\left(p_{\text{jet1}} - p_{\text{mix}}\right) + \left(p_{\text{jet2}} - p_{\text{mix}}\right)}{2} \quad 3.36$$

One problematic parameter in all these calculations is the mixing volume V_{mix} over which most the energy is dissipated. Kucher and Kind [4], Johnson and Prud'homme [52], Mahajan and Kirwan [53] or Demyanovich and Bourne [94], for example, recommend an arbitrary value of $V_{\text{mix}} = d_{\text{mix}}{}^{3}$. Schwarzer [5], for example, used the complete main duct volume as V_{mix} of the mixing nozzle investigated, but also mentions some vagueness in this choice. Comparing our experimental findings [54], the approximation of $d_{\text{mix}}{}^{3}$ for the effective mixing volume does not seem to be a bad choice. In chapter 3.3.1.3, it is shown that the mixing length in the T-mixer corresponds to $z_{\text{mix*}} \approx d_{\text{mix}}$ for the T-mixer investigated and $z_{\text{mix*}} \approx 1.5 \cdot d_{\text{mix}}$ for a Y-mixer, with the requirement that segregation has to be depleted to $99\ \%$. However, at this this point, another strategy is followed by using the tangent method in the inflection point used while evaluating the mixing times. It is postulated that most of the dissipation energy is consumed in the length that is called $z_{\text{mix*}}$ in Figure 3.41. Figure 3.44 shows the mixing lengths collected, received from such CFD "measurements". One can see that $z_{\text{mix*}} \approx d_{\text{jet}}$ for the transition and the turbulent regime. This relation is also found for numerically scaled-up mixers with a scale-up factor Z_s (handled subsequently in section 3.4.4.3, Figure 3.56) where $z_{\text{mix*,Z}} \approx d_{\text{jet}} \cdot Z_s$.

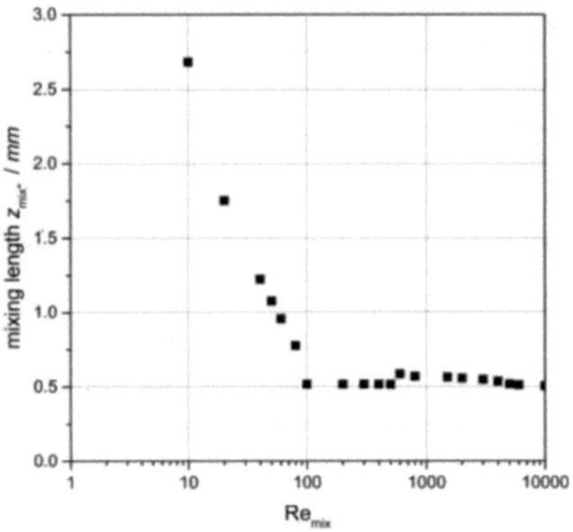

Figure 3.44: Overview of all data points extracted for z_{mix^*} at different Re_{mix} numbers for the T-following the methodology in Figure 3.41.

Mixing lengths increase for the laminar regime. Since the mixing approaches presented in the following are only valid for the turbulent regime, $V_{mix} = \pi /4 \cdot d_{mix}^2 \cdot d_{jet}$ is recommended.

Figure 3.45, left, shows the energy dissipation rates measured within CFD and the energy dissipation rates that are calculated with method 1 according to eq. 3.34 and with method 2 (eq. 3.46), with V_{mix} taking the corresponding L_{mix} from Figure 3.44.

The energy dissipations calculated for the turbulent range fit well for both methods. Clear deviations for method 1 can be observed for the laminar and transition regime. The results using the more exact method 2 for the calculation of ε, involving pressure drop and kinetic energy in the output are additionally plottedin the parity plot (Figure 3.45, right).

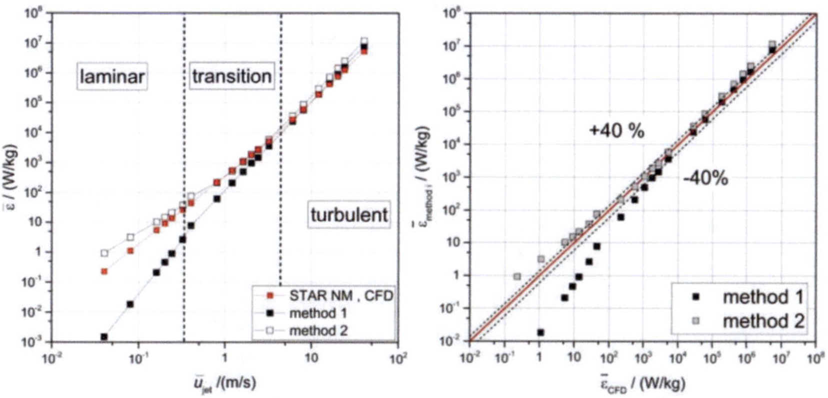

Figure 3.45: Left: Comparison of energy dissipation rates for different flow rates. Averaged CFD "measurement" in the volume V_{mix} vs. the energy dissipation rates calculated with method 1 (eq. 3.34) and method 2 (eq. 3.35) involving the pressure drop. Right: Parity plot for the two methods presented; STAR NM data taken for reference.

The Darcy-Weisbach equation and the pressure drop coefficients $\xi_i = f(\mathrm{Re}_{\mathrm{jet}})$ from section 3.3.1 are used (eq. 3.37) to calculate the specific pressure drop.

$$\Delta p = \frac{\left(\frac{\rho_{\mathrm{jet1}}\bar{u}_{\mathrm{jet1}}^{\;2}}{2}\xi_i\left(Re_{\mathrm{jet1}}\right) + \frac{\rho_{\mathrm{jet2}}\bar{u}_{\mathrm{jet2}}^{\;2}}{2}\xi_i\left(Re_{\mathrm{jet2}}\right)\right)}{2} \qquad 3.37$$

Using the relatively congruent results of method 2 compared to CFD data, ε in the laminar range can also be predicted well. Deviations, especially for low-energy dissipation values, could be linked to the sketchy experimental values of ξ_i (see Figure 3.9). However, the simpler method 1 that is presented by Johnson and Prud'homme [52] delivers almost similar results compared to values predicted by CFD in the turbulent range and should be, for this regime, a quite good approximation

Conclusions

This section gives an account of a new methodical coarse-graining approach called STAR NM, which uses spatially and temporally averaged and, furthermore, correlated CFD data. Once generated (≙ lookup chart) for an apparatus, the approach can predict mixing-influenced precipitation processes very fast (~s to min). Principal conclusions that can be extracted from this chapter are the following:

- A new indirectly coupled CFD-PBE method called STAR NM is presented which is based on the scale-separation of mixing and solid formation.
- It is shown how reasonable fast single-phase CFD "measurements" enable the prediction of experimental particle size distributions influenced by mixing very well.
- Laminar mixing phenomena and residence time effects are studied.
- It is demonstrated how time scale analysis can improve the process understanding significantly.
- Moreover, a method is presented which enables the calculation of the mixing rate determining energy dissipations in the mixing chamber from an energy balance as well as the spatial and temporal development of a mixed component fraction.

3.4.4 Classical mixing theory

In addition to the combined methods based on CFD presented in sections 3.4.2 and 3.4.3, literature provides several theoretical approaches to model mixing. The description of the mixed volume fraction α_M developing along a temporal coordinate t is one possibility available to calculate the reactant concentration effectively. Among the others are classic approaches, such as the engulfment model introduced by Baldyga and Bourne, [95, 81], and some modifications, such as the global mixing approach (GMA) [5, 96], or the extended engulfment model on various scales [80]. These models use some theoretical assumptions and rate constants. They are based on the mixing scale analysis to describe the disintegration of vortices.

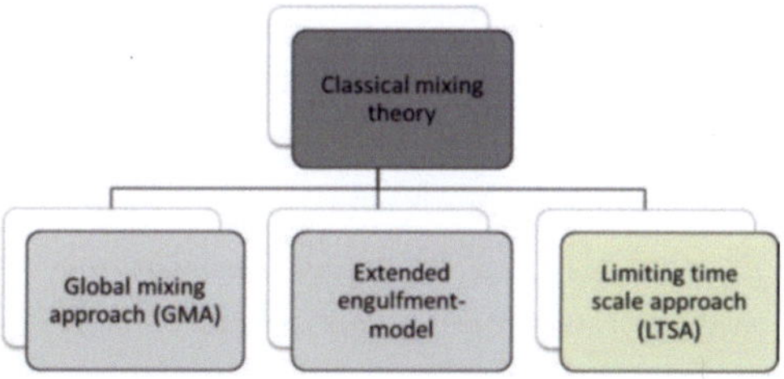

Figure 3.46: Overview of approaches which are investigated in this section. The limiting time scale approach (LTSA) approach is newly developed; the global mixing approach (GMA) and extended E-model are taken from literature and applied to the CIJMs investigated.

Figure 3.46 gives an overview of approaches that are presented and investigated in this section. Grey boxes indicate that the theory and the approaches are taken from literature and applied to the CIJM. The green box implies a new, improved model that is built up on the classical mixing theory. Predicted mixing times calculated with these models are compared to those measured within the STAR NM approach (section 3.4.3, Figure 3.41).

First of all, the theoretical background for a proper understanding of the approaches applied is presented. The basic theoretical mixing phenomena and time scales are introduced. In a further step, mixing approaches which access these time scales (e.g. GMA or LTSA, see Figure 3.46) are presented. Results shown in this section are published in Metzger and Kind [78].

3.4.4.1 Laminar time scales of mixing

According to Baldyga and Bourne [81] and Falk and Commenge [97], laminar mixing in microchannels is caused by a combination of shear and diffusion. Under certain idealized assumptions, the respective time scale of mixing can be expressed as a function of the energy dissipation rate ε. Thus, supposing a cylindrical channel with a fully developed Hagen-Poiseuille flow where $\varepsilon = (\dot{V}\Delta p)/(\rho V_{\mathrm{mix}}) = 32 \cdot \nu_{\mathrm{kin}}(\bar{u}_{\mathrm{mix}}/d)^2$ and an initial striation thickness of one half of the channel diameter, then eq. 3.38 gives the theoretical value of the laminar mixing time. Pe represents the Péclet number for mass transport, defined as the product of the Reynolds and Schmidt number ($\mathrm{Pe} = \mathrm{Re} \cdot \mathrm{Sc}$). The derivation of this equation is given in detail in [97]. Due to many idealized assumptions, these theoretical values for the laminar mixing time are too fast. Malecha et al. [98] highlight that laminar mixing is a complex interplay of stretching and folding of thin lamellae, where gradients are depleted via diffusion and which strongly depends on the mixer type used. Moreover, it is noted that the production of a maximum amount of intermaterial area between two or more initially segregated fluids in a very short time range is of high importance. Thus, in order to overcome these difficulties, Ottino et al. [99] and Falk and Commenge [97] reformulated eq. 3.38 by introducing the concept of energetic efficiency η_{mix}, see eq. 3.39. Thereby, values of η_{mix} range between 0 and 1. An extensive review of several different micromixers operated in the laminar regime [97] showed that the average value of mixing efficiency of all devices tested lies between about 3 to 5 %.

$$\tau_{diff+shear} = \frac{1}{\sqrt{2}} \left(\frac{\nu_{kin}}{\varepsilon}\right)^{1/2} \ln(1.52\ \mathrm{Pe}) \quad 3.38$$

$$\tau_{diff+shear} = \frac{d_{mix}}{8\bar{u} \cdot \eta_{mix}} \ln(1.52\ \mathrm{Pe} \cdot \eta_{mix}) \quad 3.39$$

Moreover, Falk and Commenge [97] and Commenge and Falk [100] have shown that the empirical mixing time dependency for their overview of several micromixers may be described by

$$\tau_{mix} = 0.15 \cdot \varepsilon^{-0.45} \quad 3.40$$

Figure 3.47 gives an overview of the laminar mixing times presented.

It depicts three correlations for laminar mixing from literature (see section 3.4.4.1). One is the empiric correlation which is given by Falk and Commenge [97] in their comprehensive review on micromixing (red, dashed line). Secondly, the maximal feasible mixing time for laminar mixing in a pipe flow with an initial striation thickness of half the cross-section is plotted ($\eta_{mix} = 1$) [81], (black line). Thirdly, the concept of mixing efficiency is used that charges the maximal laminar mixing time with the efficiency factor presented.

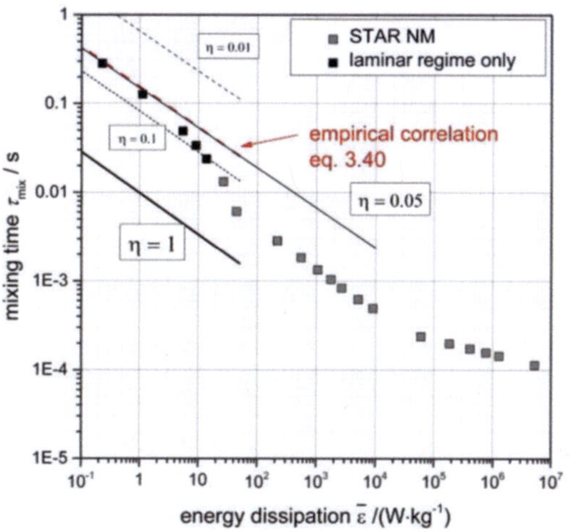

Figure 3.47: *Laminar regime: time scale analysis; in addition to the CFD mixing times measured, mixing times forecasted for different laminar mixing efficiencies* η *are added. Moreover, the empirically found correlation of Falk and Commenge [97],* $\tau_{\mathrm{mix}} = 0.15 \cdot \varepsilon^{-0.45}$ *is supplemented (red dashed line, very close to* $\eta = 0.05$*).*

Figure 3.47 shows that the mixing times extracted from CFD suit well the empirical correlation of Falk and Commenge [97] and Commenge and Falk [100] and the calculated mixing time, with a mixing efficiency of 0.05. This congruency is also given for the slope of the laminar STAR NM data that depend on energy dissipation with $\sim \varepsilon^{-0.45}$. When the transition from segregated laminar to laminar vortex flow regime occurs, mixing efficiency increases due to the important contribution of the oscillating wavy vortex plane [54].

3.4.4.2 Turbulent time scales of mixing

A spectral interpretation of mixing phenomena in turbulent flows that distinguishes certain mixing mechanisms at different turbulent subranges was developed by Baldyga and Pohorecki [101] and Baldyga and Bourne [81].

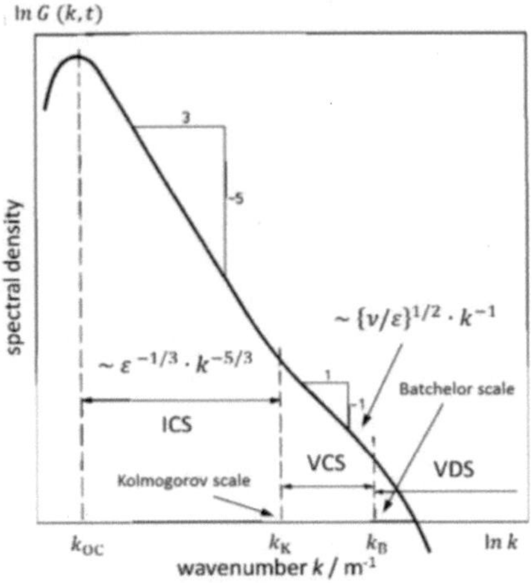

Figure 3.48: *Concentration spectrum for liquid mixtures* ($\mathrm{Sc} = 1000$)*. k represents the wave number of vortices $[k] = m^{-1}$ and is the inverse of a length scale of the vortices. (cited and modified from Baldyga and Bourne [102])*

Thereby, specific subranges with different mixing effects are identified considering the spectral density function of turbulent kinetic energy (see Figure 3.48).

Starting in the inertial convective "mesomixing" subrange (ICS; $\lambda_K < L < \Lambda_c$), distinctive regions of fluids of the initial scale Λ_c (in Figure 3.48 $\triangleq 1/k_{OC}$) are deformed and broken up by shear and elongation of the fluid. Diffusive mixing has no importance. Typical mesoscale phenomena and time scales established in literature [81, 80, 101] are the effects of turbulent dispersion, eq. 3.41, whereby a feed stream $\dot{V}_{\mathrm{jet1/jet2}}$ spreads out traverse to its local streamline and the inertial convective disintegration of large eddies, see eq. 3.42, which is responsible for the concentration variance from integral scale Λ_c to Kolmogorov scale with $\varepsilon^{-1/3}$. The integral scale Λ_c for CIJMs is chosen to be d_{jet}, and A_c is taken to a value of 1.2 [80]. One problematic value while calculating τ_s is the value of $\bar{u}$ (velocity of the flow surrounding the feed fluid [101] in the highly turbulent mixing zone). Since the dominating velocity in the mixer head is significantly higher than $\bar{u}_{\mathrm{mix}}$, we chose $\bar{u} = \bar{u}_{\mathrm{jet1/jet2}}$ in a first assumption, which surely includes some error potential, but does not influence the dependency of $\varepsilon^{-1/3}$.

turbulent dispersion
$$\tau_{\mathrm{d}} = \frac{\dot{V}_{\mathrm{jet1/jet2}}}{\bar{u} \cdot D_{\mathrm{t}}}, \quad with\ D_{\mathrm{t}} = 0.12 \cdot \bar{\varepsilon}^{1/3} \mathrm{d_{mix}}^{4/3} \tag{3.41}$$

eddy disintegration
$$\tau_{\mathrm{s}} = A_c \left(\frac{{\Lambda_c}^2}{\varepsilon}\right)^{1/3} \approx 1.2 \left(\frac{{d_{\mathrm{jet}}}^2}{\varepsilon}\right)^{1/3} \tag{3.42}$$

Mixing times according to these two mechanisms are depicted in Figure 3.49 for the T-mixer investigated. The inertial convective subrange is followed by the viscous convective subrange (VCS; $\lambda_{\mathrm{B}} < L < \lambda_{\mathrm{K}}$), where eddies are subjected to laminar strain and viscous deformation. Thin lamellar structures are formed by the stretching of the vortices. At even smaller length scales, molecular diffusion becomes important. Retarded concentration gradients below the Batchelor scale in the layered lamellae of a stretched vortex are rapidly compensated in the viscous-diffusive subrange (VDS; $L < \lambda_{\mathrm{B}}$) by molecular diffusion.

One of the most popular microscale approaches provided in literature is the engulfment model ([81] and [95]). In terms of the spectral analysis, the engulfment model, with the characteristic engulfment time τ_e, is not a true micromixing model, because it overlaps into the mesoscale range. Baldyga and Bourne [81] have shown that the shrinking lamellae in turbulent flow become embedded within stretching vortices of the scale $12\,\lambda_{\mathrm{K}}$. These vortices are stable due to the equilibrium between the work of stretching and the viscous dissipation. The incorporation of fluid by the stretching of the vortex tube is called engulfment. The engulfment time represents the lifetime of an eddy of $12\,\lambda_{\mathrm{K}}$ and depends on the energy dissipation with $\varepsilon^{-1/2}$.

engulfment
$$\tau_{\mathrm{e}} = \frac{12}{ln(2)} \left(\frac{\nu_{\mathrm{kin}}}{\varepsilon}\right)^{0.5} \approx 17.3 \left(\frac{\nu_{\mathrm{kin}}}{\varepsilon}\right)^{1/2} \tag{3.43}$$

The engulfment rate coefficient E which relates the energy dissipation to kinematic viscosity ν_{kin} reads as

$$E = \frac{1}{\tau_{\mathrm{e}}} \approx 0.058 \sqrt{\frac{\varepsilon}{\nu_{\mathrm{kin}}}} \tag{3.44}$$

In Figure 3.49, τ_e is also depicted. With regard to the dominant mechanisms, discussed further within this work, Baldyga and Pohorecki [101] have proposed a parameter M, equivalent to a Damkoehler number (eq. 3.45, see also eq. 3.33), which relates meso- and microscale time constants with respect to each other. Inertial-convective mesomixing plays the dominant role for $\mathrm{M} \gg 1$, and micromixing controls when $\mathrm{M} \ll 1$.

M-*factor*

$$\mathrm{M} = E \cdot \tau_{\mathrm{meso}} = \frac{\tau_{\mathrm{meso}}}{\tau_{\mathrm{micro}}} \qquad 3.45$$

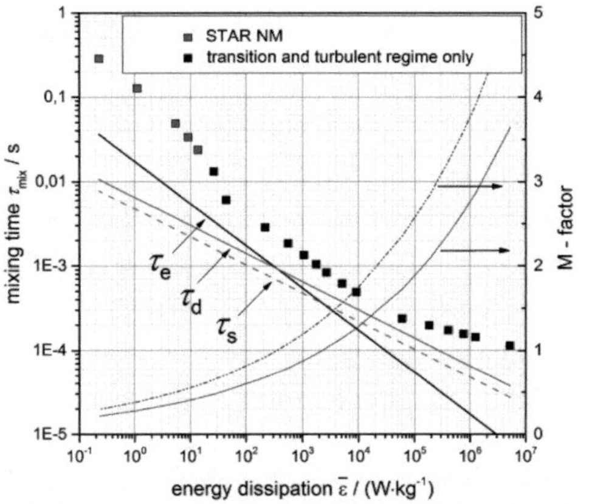

Figure 3.49: Global turbulent regime: time scale analysis including CFD data measured (single data points) and theoretical mixing times. The two M-factor trends are calculated assuming, on the one hand, τ_d and, on the other hand, τ_s as the dominant mesomixing time.

Figure 3.49 shows a comparison of turbulent time scales. The black line designates the engulfment times calculated (eq. 3.43), and the two grey lines are the mesomixing times calculated (eq. 3.41 and 3.42) for the mixer investigated that also involve geometry parameters, such as the jet diameter and the diameter of the mixing zone. Micro- and mesomixing times, more precisely the inertial convective and viscous convective subrange, differ in their exponential dependency on the energy dissipation. Therefore, the slope of data points in a double logarithmic diagram is an important indicator of the dominant mechanism. The first section in the turbulent range ($\sim 10^2 < \varepsilon < 10^4$) corresponds to the micromixing-dominated regime. The gradient of $\varepsilon^{-0.5}$

that mixing theory predicts is regained in the STAR NM data. At energy dissipations of $\varepsilon \approx 10^4 - 10^5$ W/kg, a change in slope and, thus, a change of the dominant mechanism can be observed. The inertial convective subrange gains increasing importance. This means that phenomena such as the turbulent dispersion of the feed or the inertial eddy disintegration start to be the limiting factor. This valuable information leads to interesting details that are also important for the scale-up behavior of mixing nozzles that is addressed in section 3.4.4.3. The M-factor (eq. 3.45) calculated for both mesomixing times shows where a transition of the dominant mixing mechanism is to be expected. Comparing the "measured" data to the transition forecasted, the range fits reasonable well. The differing slope of the STAR NM data in the inertial convective subrange is not fully understood. However, the calculation of mesomixing times involves certain uncertainties, such as the turbulent diffusion coefficient that is calculated following the correlation of Baldyga and Pohorecki [101].

3.4.4.3 Mixing theory-based approaches – A comparison

Baldyga and Bourne [95] showed that diffusion plays a negligible role for $\mathrm{Sc} \ll 4000$ and does not influence the reaction rate in the system. Outgoing from the engulfment-deformation-diffusion (EED) model that requires the solution of coupled, nonlinear, parabolic partial differential equations due to the diffusion terms involved, they simplified this complex set of equations to a set of ODEs (engulfment (E-) model) that neglect the influence of diffusion. The assumption that the depletion of concentration variances via diffusion plays no important role in the mixing step is justified for liquid systems, such as the model systems in our scope ($BaSO_4$-water) with $\mathrm{Sc} \approx 950$ [95]. The EED model and the basic E-model deliver the same results. All approaches investigated and presented in the following (the GMA and the E-model and its modifications, see Figure 3.46), based on second-order kinetics of mixing, are restricted to fully turbulent flows. It is assumed, considering the "measured" linear dependency for τ_{mix} of $\varepsilon^{-0.5}$ (STAR NM, Figure 3.41), that the approaches may also be valid for the transition regime (see Table 3.3). This threshold lies at $\mathrm{Re}_{\mathrm{mix}} \approx 80$ ($\bar{\varepsilon} \approx 30$ W/kg) in the CIJM investigated.

Global mixing approach

The GMA by Schwarzer and Peukert [96] and Schwarzer [5], representing an extension of the basic E-model, consists of five coupled ODEs to estimate the evolution of the mixed volume fraction α_{M}.

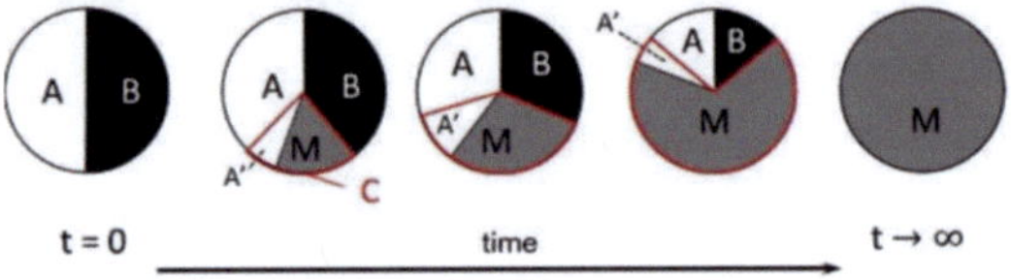

Figure 3.50: Global mixing approach: Schematic structure of the different environments in the micromixing models presented, according to Schwarzer [5].

A set of coupled ODEs must be solved to calculate the evolution of two unmixed feed streams of volume fractions α_{A} and α_{B}. The GMA reads as follows, see eq. 3.46 to 3.50:

$$\frac{\partial \alpha_{\mathrm{A}}}{\partial t} = -E\left(\alpha_{\mathrm{C}}\alpha_{\mathrm{A}} + \alpha_{\mathrm{A}}\alpha_{\mathrm{B}}\frac{\alpha_{\mathrm{A}}}{\alpha_{\mathrm{A}} + \alpha_{\mathrm{B}}}\right) \qquad 3.46$$

$$\frac{\partial \alpha_{\mathrm{B}}}{\partial t} = -E\left(\alpha_{\mathrm{C}}\alpha_{\mathrm{B}} + \alpha_{\mathrm{A}}\alpha_{\mathrm{B}}\frac{\alpha_{\mathrm{B}}}{\alpha_{\mathrm{A}} + \alpha_{\mathrm{B}}}\right) \qquad 3.47$$

$$\frac{\partial \alpha_{\mathrm{C}}}{\partial t} = E(\alpha_{\mathrm{C}}\alpha_{\mathrm{A}} + \alpha_{\mathrm{C}}\alpha_{\mathrm{B}} + \alpha_{\mathrm{A}}\alpha_{\mathrm{B}}) \qquad 3.48$$

$$\frac{\partial \alpha_{\mathrm{M}}}{\partial t} = E\left(\alpha_{\mathrm{C}}\alpha_{\mathrm{B}} + \alpha_{\mathrm{A}}\alpha_{\mathrm{B}}\frac{\alpha_{\mathrm{B}}}{\alpha_{\mathrm{A}} + \alpha_{\mathrm{B}}} + \alpha_{\mathrm{M}}\alpha_{\mathrm{A'}}\right) \qquad 3.49$$

$$\frac{\partial \alpha_{\mathrm{A'}}}{\partial t} = E\left(\alpha_{\mathrm{C}}\alpha_{\mathrm{B}} + \alpha_{\mathrm{A}}\alpha_{\mathrm{B}}\frac{\alpha_{A}}{\alpha_{\mathrm{A}} + \alpha_{\mathrm{B}}} - \alpha_{\mathrm{M}}\alpha_{\mathrm{A'}}\right) \qquad 3.50$$

As explained by Schwarzer [5] and Schwarzer and Peukert [96], the closure of the ODEs reads as $\alpha_{\mathrm{A}} + \alpha_{\mathrm{B}} + \alpha_{\mathrm{C}} = 1$ and $\alpha_{\mathrm{A'}} + \alpha_{\mathrm{M}} = \alpha_{\mathrm{C}}$ and does not have to be solved. It can, therefore, be taken to prove the correctness of the solution. The initial values are chosen to be $\alpha_{\mathrm{A}}(t = 0) = \alpha_{\mathrm{B}}(t = 0) = 0.5$; $\alpha_{\mathrm{C}}(t = 0) = \alpha_{\mathrm{A'}}(t = 0) = \alpha_{\mathrm{M}}(t = 0) = 0$.

One unclear deficit of the GMA model is the existence of zone A', which represents the unmixed solution of component A in the contact zone C, and which is not implemented for component B. It represents a dead time element of one eddy turnover. However, for the purpose of comparability, this approach is implemented according to Schwarzer [5], who achieved relatively congruent results comparing experiments and simulated PSD. Figure 3.51, left, shows the evolution of the volume fractions involved solving the set of ODEs of the GMA. Figure 3.50 and the associated set of equations (eq. 3.46 to 3.50) help one to understand the specific evolutions. In terms of comparison to STAR NM data (see section 3.4.3), where the supersaturation buildup in the CIJM is "measured", corresponding continuing calculations are also executed for the classic mixing approaches.

The calculation of the "mixed," molar concentrations of Ba^{2+} and SO_4^{2-} out of the mixed volume fraction α_M and the use of the activity coefficient model (see section 2.2.3) leads to the supersaturation buildup in Figure 3.51, right. Characteristic mixing times are evaluated using the same tangent method presented in Figure 3.41, left.

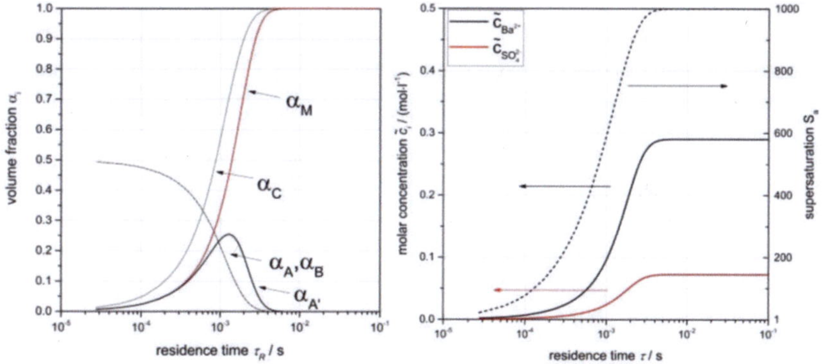

Figure 3.51: Left: Progress of the volume fractions α_i for $\bar{\varepsilon} = 1000$, solving the set of ODEs for the GMA. Initial values for α_A and α_B are chosen to be 0.5, right: Concentration and supersaturation evolution calculated for $S_{a,nom} = 1000$ and $\mathrm{R_c} = 5$ from the mixture volume fraction using the GMA.

As one can see, the GMA involves a dead time element A' that shifts the evolution of α_C (mixed region, neglecting $\alpha_{\mathrm{A}'}$) for one eddy turnover time to longer mixing times (α_M). The need for this decelerating element is not fully understood. Therefore, no such intermediate zone is considered in a new formulation of the mixing model which is presented in the following and which is called the “limiting time scale approach” (LTSA).

Extended engulfment model (mixing on various scales)

An extended formulation of the engulfment model [80] takes mixing on various scales into account. In addition to micromixing, inertial convective mixing by disintegration of large eddies (see eq. 3.42) is considered. The basic idea of the approach is the existence of partially segregated “islands” (see Figure 3.52). Inwardly, micromixing occurs along the engulfment model of mixing including self-engulfment. α_{iu} (i = component *A* and *B*) is the volume fraction of micromixed fluid within these islands. Mixing is promoted by the engulfment of the unmixed volume fraction of component i and moderated by the engulfment of the volume already mixed with the unmixed regions (self-engulfment). Additionally, these “islands” embedded in a “sea” that does not yet contain component i must be mesomixed by the mesomixing rate constant $1/\tau_s$. α_i relates to the micromixed volume fraction in the entire control volume. Consonant with Baldyga et al. [80], the governing equations, adapted to the two-feed situation ($\alpha_A + \alpha_B = 1$) in a CIJM, read as

$$\frac{\partial \alpha_{\mathrm{A}}}{\partial t} = -E\alpha_{\mathrm{A}}\left(1 - \frac{\alpha_{\mathrm{A}}}{\alpha_{\mathrm{Au}}}\right) \tag{3.51}$$

$$\frac{\partial \alpha_{\mathrm{B}}}{\partial t} = -E\alpha_{\mathrm{B}}\left(1 - \frac{\alpha_{\mathrm{B}}}{\alpha_{\mathrm{Bu}}}\right) \tag{3.52}$$

$$\frac{\partial \alpha_{\mathrm{M}}}{\partial t} = E\left(\alpha_{\mathrm{A}}\left(1 - \frac{\alpha_{\mathrm{A}}}{\alpha_{\mathrm{Au}}}\right) + \alpha_{\mathrm{B}}\left(1 - \frac{\alpha_{\mathrm{B}}}{\alpha_{\mathrm{Bu}}}\right)\right) \tag{3.53}$$

$$\alpha_{iu} = \frac{\alpha_{i0}}{\alpha_{i0} + (1 - \alpha_{i0}) \cdot e^{(-t/\tau_s)}} \tag{3.54}$$

When only micromixing controls ($\alpha_{i,u} = 1$), the extended engulfment model reduces to the engulfment model of mixing with self-engulfment. Calculations are executed with the initial volume fractions $\alpha_{\mathrm{A0}} = \alpha_{\mathrm{B0}} = 0.5$.

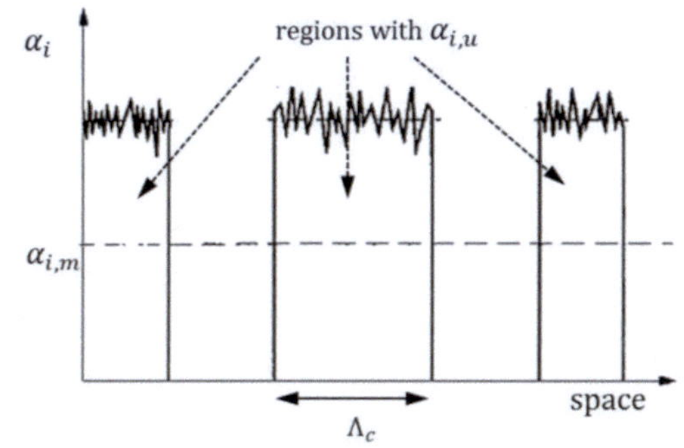

Figure 3.52: Basic idea of the extended engulfment model with mixing on various scales (adapted and modified from Baldyga et al. [80]).

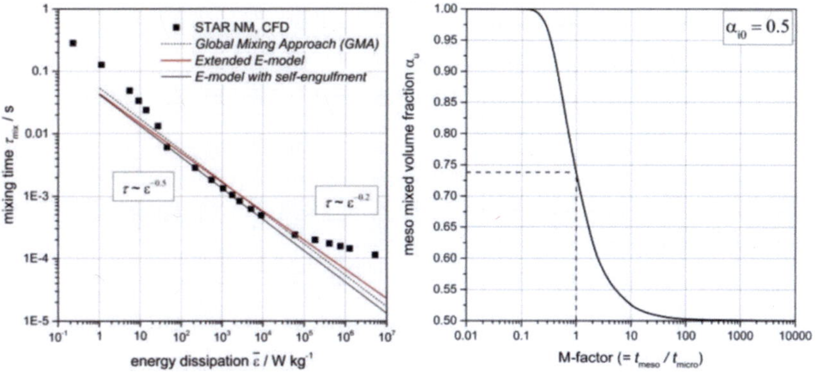

Figure 3.53: Left: Predicted mixing times: A comparison between "measured" STAR NM results (black squares) and classic approaches, mixing times from the extended E-model involving the mesoscale time τ_s (red line), standard E-model with self-engulfment ($\alpha_{iu} = 1$)(black line) and GMA (dashed black line); right: Influence of the time ratio t_{meso}/t_{micro} on the volume fraction α_{iu} (eq. 3.54) in the extended E-model exemplarily with $\alpha_{i0} = 0.5$.

Calculations such as the ones presented for the GMA are executed using the extended E-model (eq. 3.51 to 3.54) that involve, in addition to micromixing, the inertial convective mixing step, implemented over the volume fraction α_{iu}. To check the plausibility of the model, α_{iu} is set to a value of 1. As described by Baldyga et al. [80], in this case, the extended E-model reduces to the engulfment model of mixing with self-engulfment.

Figure 3.53, left, shows the mixing times calculated for the GMA (section before) for the extended E-model and for the standard E-model with self-engulfment ($\alpha_{iu} = 1$). Thereby, two important basic facts can be determined. Firstly, one can assert that all mixing times calculated with the mixing approaches tested lie in the similar range the STAR NM approach predicts when micromixing by engulfment dominates (see Figure 3.49). Moreover, the slope of the GMA approach and the engulfment model with self-engulfment fits the slope of the STAR NM reference data very well. Actually, the standard E-model with self-engulfment lies exactly on the mixing times calculated within CFD. The time shift the GMA involves considering a compartment A' cannot be observed. As described in section 3.4.4.2, a change in slope is observed that is interpreted as a change of the dominant mixing mechanism (for $\varepsilon > 10^4 - 10^5\ W/kg$). The inertial convective subrange starts to limit the mixing velocity. The extended E-model is a model that accounts for the two mixing scales and this phenomenon. The red line in

Figure 3.53, left, shows the mixing times forecasted from the extended E-model. The consideration of the eddy disintegration step in this formulation leads to a change in slope we cannot observe in the CFD data.

Figure 3.53, right, shows the influence of the M-factor (see eq. 3.45) on the mesomixing contribution value α_{iu} in the extended E-model (by substitution of t/τ_s by $1/M$ in eq. 3.54) for an exemplarily chosen $\alpha_0 = 0.5$. This figure helps to interpret the mixing times calculated and the change in slope observed. Mesomixing plays no role for an $\alpha_{iu} = 1$. This value is derived firstly for an M-factor of 0.1, that means when $t_{\text{micro}} = 10 \cdot t_{\text{meso}}$. This is only the case for very low-energy dissipation values (see Figure 3.49, right). Following this mixing approach, all values that lie above an M-value of 0.1 are affected and decelerated by mesomixing phenomena and lead to a sinking slope. A clear inertial convective limitation is predicted for an M-factor of 1. Mixing times deviate with increasing energy dissipation more and more from the "measured" STAR NM data.

New mixing law formulation – Limiting time scale approach

To lower the deviations in mixing time, especially for high energy dissipation rates ($\varepsilon > 10^4\ W/kg$) discussed in the section above, a new mixing approach is proposed that has its origins in the classic engulfment model which fits the STAR NM data very well, especially in the transition regime. Instead of the contribution of mesomixing the extended E-model provides, and considering data from STAR NM mixing time "measurements," a set of ODEs is proposed which differs. Thereby, the mixing rate coefficient, considering the three dominant mechanisms: engulfment, turbulent dispersion and eddy disintegration, is set to be the minimal value of all three coefficients as the rate-determining step. This procedure leads to an extension of the turbulent mixing model (E-model with self-engulfment) which includes the initial convective and viscous convective subrange. Figure 3.55 shows the two approaches. One can see a characteristic bend at $\bar{\varepsilon} \approx 8 \cdot 10^3\ W/kg$, when mesomixing becomes rate-determining.

In addition to mixing on the microscale (engulfment), this model, in a similar way to the extended E-model, considers limitations from the inertial convective subrange, but in a different manner. Basic assumptions of the approach with second-order kinetics are the same as those used for the E-model, including self-engulfment. This means that mixing takes place by the exchange of material of zone A and B, as well as fresh feed (zone A, respectively B) with the fluid mixed already (zone M) (see Figure 3.54). Additionally, the concept of the approach is to evaluate the limiting time scale as the decelerating mixing step. The mixing phenomena engulfment, eddy disintegration and turbulent dispersion are considered as competing mechanisms of mixing. Thus, the limiting process-dominating mixing constant is calculated for the specified flow conditions (see eq. 3.58). Inertial convective subrange phenomena, such as turbulent dispersion or the eddy disintegration, can pose the limiting step even for high flow rates, as discussed later [80]. Since only turbulent mixing phenomena are considered, the approach should only be valid for turbulent flow. The ODE formulation is set up based on the work of Baldyga and Bourne [81] and compared with mixing times predicted by the STAR NM approach. The mixing formulation reduces to the E-model of mixing with self-engulfment for $k_{\text{mix}} = E$.

$$\frac{\partial \alpha_A}{\partial t} = -k_{\text{mix}}(\alpha_\text{A}\alpha_\text{B} + \alpha_\text{A}\alpha_\text{M}) \qquad 3.55$$

$$\frac{\partial \alpha_\text{B}}{\partial t} = -k_{mix}(\alpha_\text{A}\alpha_\text{B} + \alpha_\text{B}\alpha_\text{M}) \qquad 3.56$$

$$\frac{\partial \alpha_\text{M}}{\partial t} = k_{\text{mix}}(2 \cdot \alpha_\text{A}\alpha_\text{B} + \alpha_\text{M}(\alpha_\text{A} + \alpha_\text{B})) \qquad 3.57$$

$$k_{\text{mix}} = \min(E, k_{\text{meso,d}}, k_{\text{meso,s}}) \qquad 3.58$$

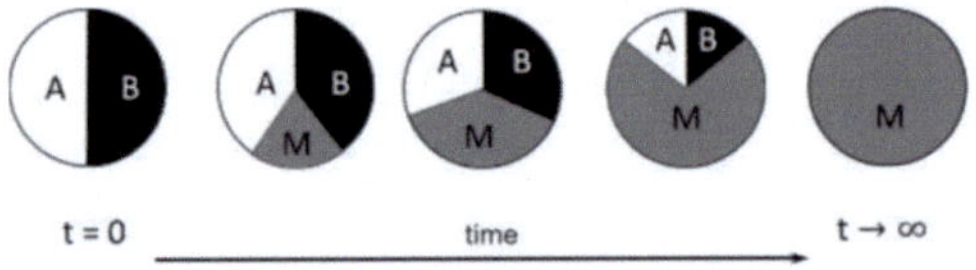

Figure 3.54: New formulation of the mixing approach. Graphical illustration adapted to Figure 3.50.

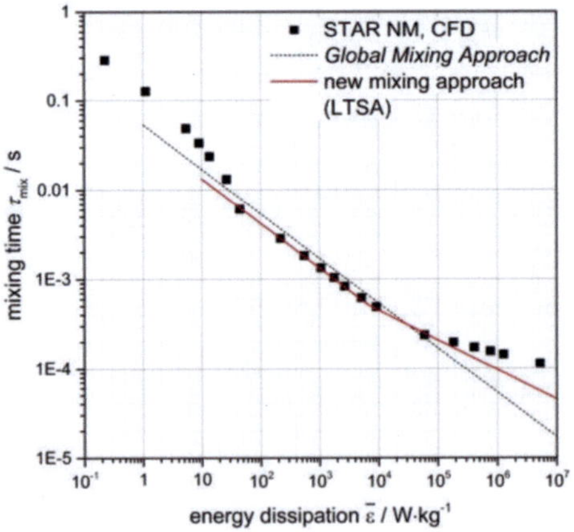

Figure 3.55: Comparison of the GMA and the new formulation (LTSA) in the context of the measured mixing times within CFD.

Scale-up of CIJMs – A hypothesis

The formulation of the LTSA leads to a new hypothesis concerning the scale-up behavior of CIJM. As mesomixing time is dependent on geometry parameters, we postulate that upscaling is also possible with the STAR NM approach. By multiplying the geometry parameters ($d_{\mathrm{mix}} \cdot Z_s;\ d_{\mathrm{jet}} \cdot Z_s$) with a scale-up factor Z_s, the lines for $Z_s = 5, 10$ and 50 are calculated according to the LTSA approach in the section before (see Figure 3.55).

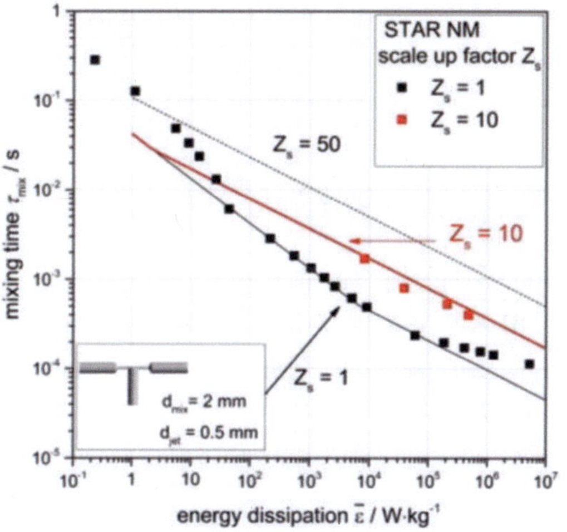

Figure 3.56: Scale-up hypothesis derived from the new formulation of the mixing model (LTSE) considering micro- and mesoscale. The red squares show STAR NM data for a scaled-up mixer with a scale-up factor of $Z_s = 10$.

Calculations predict that the inertial convective subrange, which takes into account turbulent dispersion of the feed or the initial disintegration of large eddies, requires more and more time with increasing jet diameter and mixing chamber diameter and, finally, becomes the rate-determining step for large CIJMs. With an increasing jet diameter, the inertial blob of fluid that must be reduced down to microscale by mesoscale rates requires more time. Such observations have been reported by Baldyga and Pohorecki [101] and Johnson and Prud'homme [52]. Marchisio et al. [103] "measured" similar characteristic bends in slope comparing two differently sized CIJMs while plotting the mixing time evaluated within CFD over the $\mathrm{Re_{jet}}$ number. However, they could not clearly classify their results, but remarked that contributions of micro- and mesomixing lead to these effects.

The STAR NM calculations for a scaled-up mixer with $Z_s = 10$ support our hypothesis strongly. The calculations presented address scale-up problems for CIJMs which are well-known. They can explain, for instance, that larger CIJMs, even if the same mean energy dissipation is adjusted, will produce larger particle sizes due to a slower supersaturation buildup which is induced by the slower reactant mixing. Possibly, the LTSA also explains the small geometric correction that is recommended by Johnson and Prud'homme [52] for the mixing time estimation when scaling-up CIJMs with a constant jet inlet velocity. Moreover, these corrections could be disposable with the approach presented.

Conclusions

The section presented basically builds up on coarse-graining data from the newly developed STAR NM approach which enables the exposure of unique mixing time-scale information. The chapter sets these CFD-predicted times in comparison to theoretically calculated mixing times and shows some remarkable points which will be summarized in the following.

- Mixing models from literature (engulfment model, GMA) are adapted and applied to the T-mixer investigated. Highly congruent results in mixing times measured in CFD and mixing times predicted by mixing theory are found for both the laminar and the turbulent regime.
- An interesting point, detected by CFD, is the enhanced influence of the inertial convective subrange (mesomixing effects) at high-energy dissipation rates. This effect is not taken into account by classic mixing models, such as the engulfment model of mixing or the GMA, which only consider micromixing by engulfment. Based on these findings, a new formulation of a modified engulfment model is proposed using a set of coupled ODEs with a second-order kinetic that takes micro- and mesoscale in the turbulent regime into account. In the following, the approach is called the LTSA.
- This new model allows, as further consequence, the estimation of scale-up effects in CIJMs, since mesomixing depends on the apparatus geometry. The STAR NM data for a scaled-up CIJM support this theory, which will be further promoted in section 4.4.

4 CIJM – Flow sheet simulation and approach potential

This chapter combines methods and models that have been proposed and developed in chapters 2 and 3 to set up a fast, predictive flow sheet model for the process simulation of precipitation processes in CIJMs. Thereby, with the module presented, the emphasis is on creating a high degree of flexibility concerning the parameter space addressed in section 1.

The aspects "influence of supersaturation," "influence of mixing predicted by the new theoretic mixing law formulation LTSA," "the transferability to other materials" and the topic "scale-up" are addressed.

The newly developed transient simulation tool "DYnamic Simulation of SOLids Processes – Dyssol" [104] is used for the investigations of such influences. The CIJM module has been implemented within the C++ framework and uses the Dyssol infrastructure. An explanation is given in the following section.

4.1 Flow sheet simulation

Flow sheet simulations are executed using the newly developed flowsheet environment *Dyssol* (based on C++) as a framework system [104]. The CIJM has been implemented within the software as a self-contained module which can be initialized and where all input data and user-dependent parameters can be allocated via the *Dyssol* user interface. The basic program structure that has been created and chosen for the simulation is shown in Figure 4.1.

Thereby, a beneficial interplay between *C++* (*Dyssol*), the HRFVM solver for the population balance in *Matlab* (*Matlab*-engine), and the hydrochemistry software *PhreeqC* is realized. Mixing approaches, for example, are integrated into the section "population balance/component balances" as described in section 3.4.3. The flexibility of the structure is tested in the following.

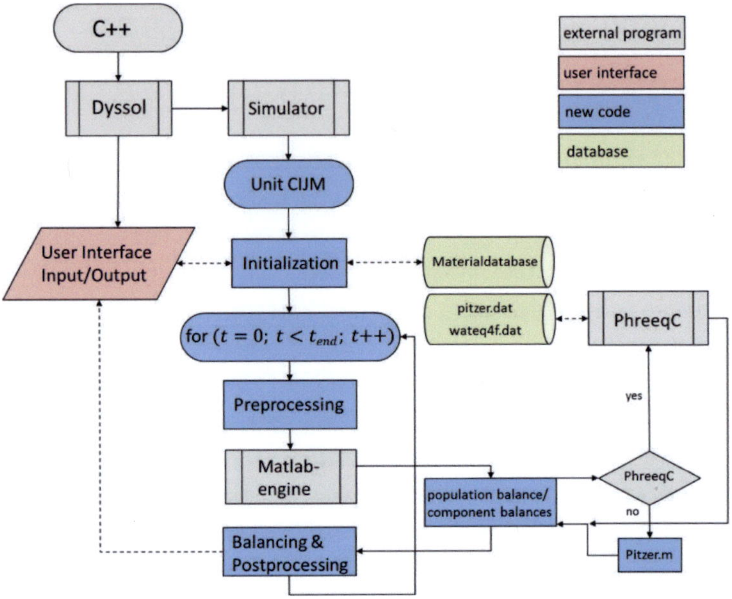

Figure 4.1: General procedure of the flowsheet simulation with Dyssol for confined impinging jet mixers (CIJMs). Thereby, the synergy of C++, Matlab engine and PhreeqC is used.

4.2 The influence of mixing using the limiting time scale approach

In this section, the influence of mixing predicted by the new scaling capable and flow sheet suitable mixing law formulation (LTSA, see section 3.4.4.3) is investigated within *Dyssol*. As shown in section 3.4.2 and 3.4.3, CFD methods which directly or indirectly couple flow mechanics and PBEs are suitable tools to investigate this influence and decode dominant procedures. Analytic approaches such as the LTSA (section 3.4.4.3) may be more efficient to set up a fast (~seconds to minutes) and flexible flow sheet tool.

Simulation is performed calculating the energy dissipation by method 2, proposed in section 3.4.3.7. Population balancing is executed according to Gunawan et al. [12] and LeVeque [9] (see 3.4.3.3), using the high resolution finite volume method presented (see section 2.1.1). The new mixing law formulation LTSA (see section 3.4.4.3) and the STAR NM method are embedded. Instead of an initially mixed system at $t = 0$ (dashed line in Figure 4.4, left), the buildup of supersaturation as the driving force of nucleation and growth due to mixing by the new mixing law formulation can be considered (see black line in Figure 4.4, left). Thereby, depending on the flow conditions, the inertial convective or viscous convective time scale is predominant. Simulated PSDs are then affected by the superposed mechanisms mixing and supersaturation depletion due to particle formation and growth. Secondary mechanisms of particle formation, such as agglomeration and aggregation, are not considered, since the primary particles of $BaSO_4$ are supposed to be electrostatically stabilized for a free lattice ion ratio of $\mathrm{R_c} = 5$ (excess of barium ions) (see section 2.4).

Predictions investigated by the coupling of LTSA and the population balance solver (chapter 2.1.1) for the mean particle size $L^P_{50,0}$ (see Figure 4.2) enable a satisfying accuracy for the T-mixer. All simulation results are extracted from the *Dyssol* user interface output.

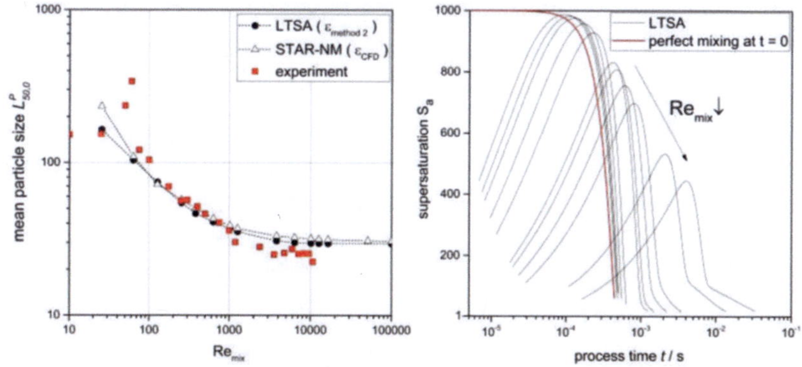

Figure 4.2: The influence of mixing calculated with the flow sheet simulation tool presented in Figure 4.1. Left: Mean particle sizes measured compared to those calculated with the STAR NM method and the LTSA method. Right: Supersaturation curves for the different flow rates using the LTSA approach.

4.3 The influence of supersaturation

The nominal supersaturation which is adjusted by the reactant ratio is one key parameter influencing the crystal mass precipitated and the resulting PSD. Model calculations have been performed for the model system $BaSO_4$ to validate the flowsheet. Kügler and Kind [3] give a robust experimental database for the primary particle sizes precipitated at varied nominal supersaturations (see Figure 2.11). By adding a surfactant (*Melpers 0045*, see section 2.4), they have been able to suppress aggregation and dendritic growth for $S_{a,\mathrm{nom}} < 600$, which is one problematic aspect regarding the model system $BaSO_4$. The flowrates for the experimental data have been adjusted adequately high ($\mathrm{Re}_{\mathrm{mix}} > 3000$) to neglect mixing phenomena.

Figure 4.3 shows the simulation results for two different interfacial tensions (see also Figure 2.11). Kügler et al. [3] used the activity coefficient approach consonant with Bromley for their experimental setup. Since this work uses the approach of Pitzer, all data are converted into the corresponding activity-based supersaturation.

On the one hand, calculations are performed for an interfacial energy $\gamma^{SL} = 0.1284\, J/m^2$, which is the value calculated according to eq. 2.41 proposed by Mersmann [105]. Moreover, $\gamma^{SL} = 0.1181\, J/m^2$ is the value presented by Kucher et al. [4] and Schwarzer [5] for $S_{a,\mathrm{Bromley}} = 1000$ and $\mathrm{R_c} = 5$ ($\triangleq$ $S_{a,\mathrm{Pitzer}} = 813$ $\mathrm{R_c} = 3.7$) which takes the adsorption of ions and the accompanied lowering of the interfacial energy into account.

In their simulations, Kügler et al. [3] and Schwarzer [5] consider this adsorbing effect which strengthens for the higher ion concentrations involved and which leads to a blend between the two simulated lines in Figure 4.3 for increasing supersaturations. Since this behavior is a $BaSO_4$ characteristic for the general flowsheet approach with higher flexibilities, it is not considered. However, the influence of supersaturation can be reproduced well with the approach presented.

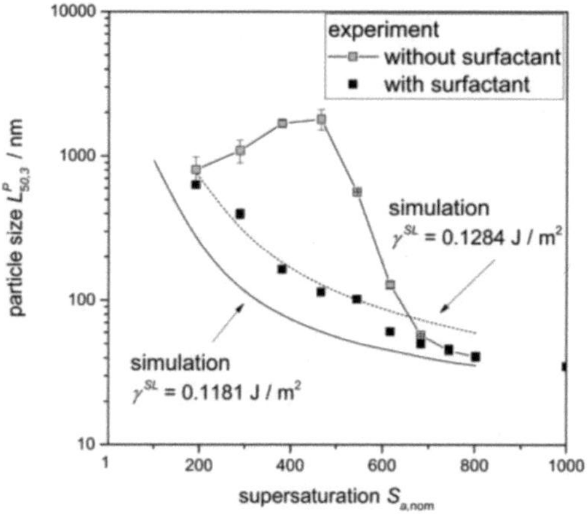

Figure 4.3: Comparison of mean volumetric particle sizes $L^P_{50,3}$ measured for $BaSO_4$ with and without surfactant (Melpers 0045, see section 2.4) and simulated particle sizes using the general flow sheet approach and the kinetics presented in section 2.3.1 and 2.3.2. Calculations are executed considering two different interfacial energies. $\gamma^{SL} = 0.1284\,J/m^2$ is the value calculated according to eq. 2.41, $\gamma^{SL} = 0.1181\,J/m^2$ is the adapted value proposed by Kucher et al. [4] and Schwarzer [5]. Experimental data are taken from Kügler et al. [3].

4.4 Scale up of CIJMs

The new mixing law formulation is developed comparing numerically measured mixing times in a T-mixer with the characteristic dimensions $d_{\text{mix}} = 2\,mm$ and $d_{\text{jet}} = 0.5\,mm$. Additional tests with two further CIJMs in Y-configuration have been performed to extend the importance of the approach and to show its application for fast reacting systems, such as precipitation (see Table 1). Experimental data of precipitated $BaSO_4$ particles as a model system are obtained at $S_{a,\text{nom}} = 1000$ and a free lattice ion ratio of $\text{R}_\text{c} = 5$ ($\tilde{c}_{Ba^{2+},\text{nom}} = 0.2902\,mol\,/l$, $\tilde{c}_{SO_4^{2-},\text{nom}} = 0.0723\,mol/l$) using the educts Na_2SO_4 and $BaCl_2$ and the activity model of Pitzer [23] (see chapter 2.2.3). The characteristic geometry data of the mixers are shown in Table 4.1. The T-mixer presented in Figure 3.1, which is intensively investigated in the sections ahead, is also listed for comparison.

Table 4.1: Characteristic dimensions of the mixers investigated. The T- and Y-mixer geometry with $d_{mix} = 2\ mm$, is presented in Figure 3.1. The enlarged Y-mixer is constructed in geometrical analogy.

parameter	T-mixer	Y- mixer 1	Y-mixer 2
jet diameter d_{jet} / mm	0.5	0.5	1
jet length L_{jet} / mm (= end of diminishing pipe to the impinging point, see Figure 3.1)	3	3	3
mixing zone diameter d_{mix} / mm	2	2	3
jet angle	180°	150°	150°

Figure 4.4, right, presents the results of mean particle sizes $L^P_{50,0}$ received by either experiment or population balancing for different flow rates.

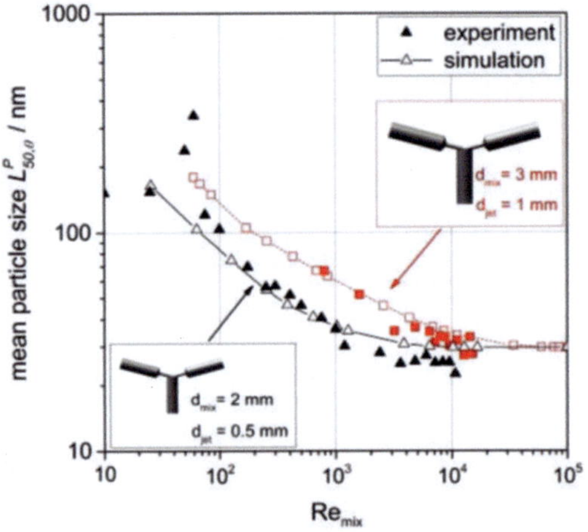

Figure 4.4: Simulated mean particle sizes $L^P_{50,0}$ for two different Y-mixers using the new mixing law formulation at different flow rates.

Three important conclusions concerning the influence of mixing can be drawn from Figure 4.4, right. Firstly, in addition to the T-mixer investigated, the mixing concept and the calculation of the adequate energy dissipation by the mean jet velocity also works quite well for the CIJM in a Y configuration. The concept of effective mixing length L_{mix} (see section 3.4.3.7) should be the most problematic factor when describing differently angled devices, since the mixing length may increase with a steeper jet angle [54]. $L_{mix} = d_{jet}$ is chosen for the simulations presented, as explained in section 3.4.3.7. Secondly, the behavior of the enlarged Y-mixer with bigger jets is very detailed, which gives, in addition to the numerical evidence which is shown in Figure 3.56, another suggestion of a scaling concept that works well. Experimental investigations on larger mixing nozzles than those investigated so far should be performed in the future to draw comprehensive conclusions. Thirdly, mixing in the laminar regime delivers deviating results, since the new mixing formulation is inherently valid only for the turbulent regime. Additional deviations may arise in terms of the effective mixing length (see Figure 3.44) and in calculating the corresponding energy dissipation (Figure 3.45).

4.5 Transferability to other material systems

The expansion of the CIJM module presented by the hydrochemistry software *PhreeqC* leads to a high degree of freedom when choosing inorganic precipitating model systems, since activity coefficient calculations are possible for almost every combination of ions. The only limitation the module provides at this stage of development is the fact that only one solid phase is considered within the population balancing. Different single salts can be precipitated easily or, for example, the ionic strength can be changed. Solubility data are directly extracted from the *PhreeqC* material database, but can also be changed manually in case more accurate data are available in literature. Figure 4.5 shows exemplarily a comparison between experimentally available data for the system SrF_2 and simulation results. The Wateq model (see section 2.2.3) has been used for the simulations since no Pitzer parameters are listed in literature. The interfacial tension is calculated using eq. 2.41.

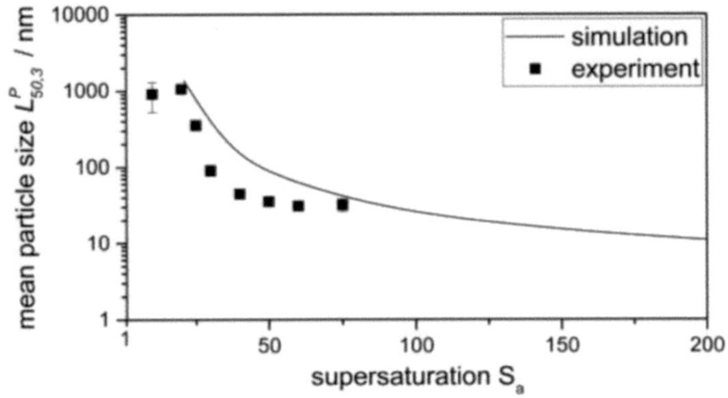

Figure 4.5: Precipitation of strontium fluoride SrF_2, comparison of flowsheet model data to experimental ones. Experimental data taken from Ailinger [106].

Figure 4.6: SEM image of SrF_2; from left to right $S_a = 10, 30$ and 75 at $R_c = 0.5$. SEM image are taken from Ailinger [106].

It can be shown that the simulation can predict the experimental results fairly well. If precise experimental data of other material system exist, further validations can be made easily.

4.6 Interconnections – consecutive combined CIJMs

A hypothetical scenario of two CIJMs is tested to check the module flexibility in terms of interconnectivity. This section is a preliminary study for the loop calculations performed for the ST reactor in section 4. As shown in Figure 4.7, the PSD created initially in mixer 1 is handed over into a second CIJM. Thereby, a barium sulfate surplus in mixer 1 is adjusted. An additional sulfate stream mixed with all the rest of the barium ions in the second mixer leads again to a supersaturation which is now created under almost stoichiometical conditions ($R_{c,2} = 0.7$).

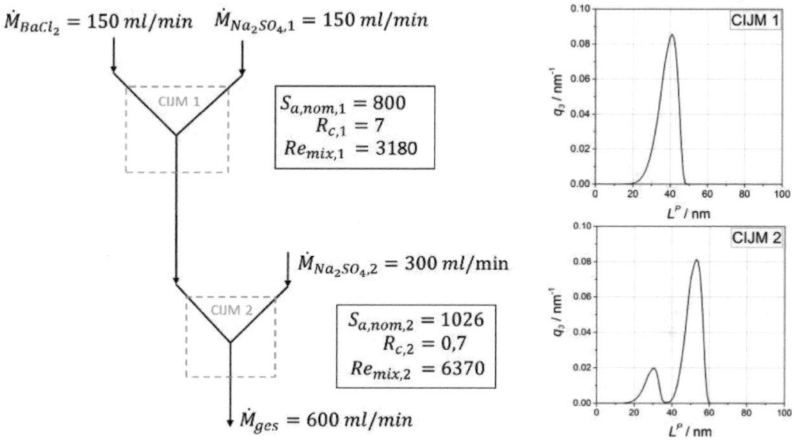

Figure 4.7: Series of two interconnected mixers, module check and visualization of the interplay between nucleation and growth.

Results in Figure 4.7 show that particles formed in mixer 1 grow subsequently, whereas additional crystals are created newly. Thus, the module is capable of handling particle-loaded reactant streams which are considered within the primary mechanisms nucleation and growth.

4.7 Computational time

In a further step, calculation times are estimated for the flow sheet module. A reasonable computational effort while predicting PSDs within the transient flow sheet simulation tool is one necessary key feature.

Figure 4.8 shows some specific details for the models used. It can be shown that an increase in the class number (a), as well as the decrease in supersaturation (b) enlarges calculation times strongly. The coupling to the PhreeqC software (c) also leads to an enhanced effort. Since the PB algorithm and the software-coupling are realized within an in-house code, some improvements in the computational time might be possible. The times required, shown in Figure 4.8, are quite satisfactory for the purpose needed.

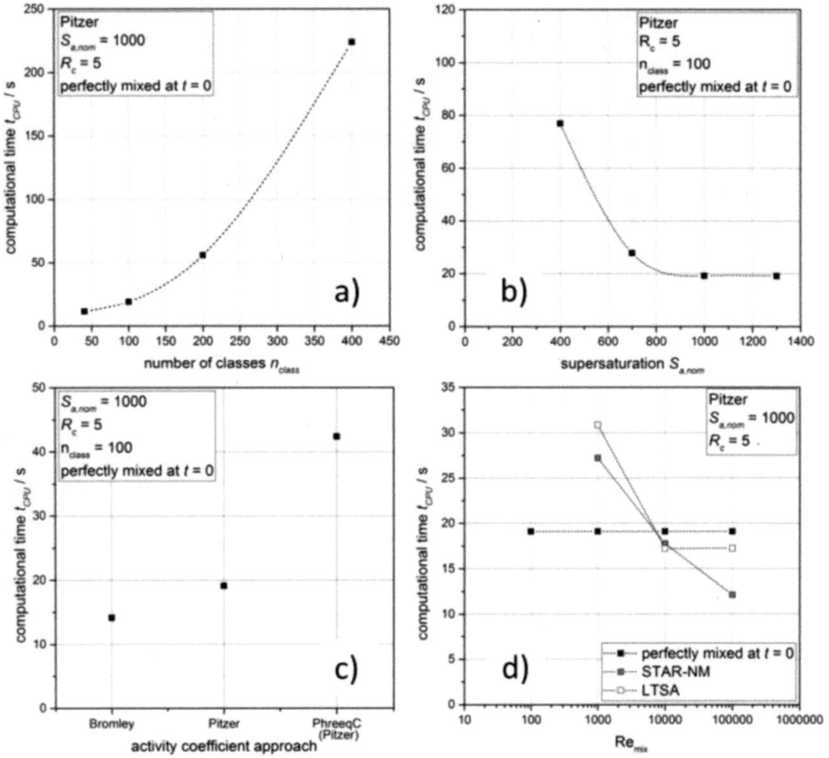

Figure 4.8: Overview of simulation times within the Dyssol-CIJM module. Constantly held parameters are specified in the legend. a) Influence of the number of classes used in the HRFVM solver, b) influence of the nominal supersaturation, c) influence of the activity coefficient approach used – additionally, the influence when coupling to the software PhreeqC is added, d) influence of the mixing model used. The LTSA equations are solved with an ODE45 solver in MATLAB.

5 Method transfer – stirred-tank reactors

This chapter, in a further step, seeks to transfer the methods which are developed and presented in chapters 3 and 1 to more complex processes such as precipitation in ST reactors. Although ST reactors are probably one of the best examined apparatuses in process engineering, highly spatially and temporally restricted processes occurring during precipitation in a complex inhomogeneous flow field brings up important open research questions. The multiscale character of precipitation mechanisms in the range of milliseconds to seconds and global process times of minutes to hours is a challenging aspect for the process simulation. Regarding the scale-structure in Figure 1.1, pursuing ideas on how to handle such a gap in time are necessary.

The chapter picks up methods such as the STAR NM addressed in the previous chapters to connect micro- and mesoscale in continuous ST reactors (CSTR)s. It has been demonstrated how the successful interplay of CFD and PB enables the prediction of particle sizes generated in the benchmark apparatus CIJM. These methodical advancements are transferred to ST reactors. Hereby, multicompartment approaches seem to be an adequate solution to tackle existing deficits in process simulation on a macro- and process-equipment scale. An appropriate flowsheet model is presented to approach this topic by creating an experimental and numerical framework.

5.1 Aim and scope

Stirred-tank reactors are probably the apparatuses most commonly used for precipitation processes in industry [1, 107]. Thereby, the flow field in a ST reactor is exposed to a highly nonuniform input of energy. Flow velocities in different tank zones differ distinctly. Figure 5.1 shows the wide variety of energy dissipation rates ε as a quantity of mixing intensity (see section 3) in a ST reactor using a Rushton turbine. Consequently, the predictive description of solid formation processes by precipitation which may be strongly influenced by these local energy dissipation rates and the specific flow field (see section 3) pose an enormous challenge.

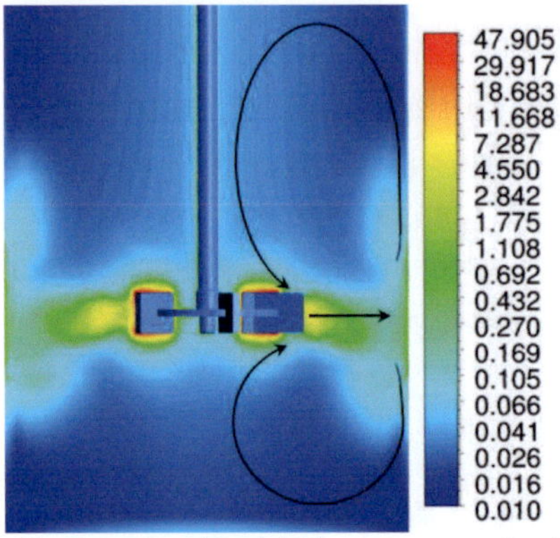

Figure 5.1: Energy dissipation in a stirred-tank (ST) reactor according to DIN 28131 [108] ($V_{ST} = 11l, n = 300\ rpm$) – snapshot from CFD with absolute values of ε in W/kg.

Considering, for example, a semi-batch process with a feed time of several minutes to hours, crystalline particles that are formed initially circle many residence times within the flow field. Some crystals may pass the supersaturated reaction zone from time to time and grow, some stay longer in the bulk, where no supersaturation exists. Additionally, new nuclei are formed all along the process time in the reaction zone situated close to the feed-pipe. This spatial and temporal multiscale behavior of the particle formation step (~ms) and the specific process characteristics (~min-h) requires hybrid simulation approaches, since fully resolved CFD simulations are not suitable to cover minutes to hours of process times. This section tries to elaborate on how the scale-decoupling methodology (e.g. STAR NM) presented in the previous sections may help to simulate processes even more complex than the CIJM. The central hypotheses that are investigated in this section read as follows

1) *Stirred-tank reactors can be abstracted by an equivalent circuit of at least two interconnected compartments. This basic consideration must be urgently validated experimentally before an appropriate model can be set up.*

2) *CFD "measurements" (STAR NM) represent the missing piece of the puzzle to identify process dominating regions and to extract adequately averaged supersaturations. Furthermore, important mass flow rates which are passing through the supersaturated region and, thus, which are participating in the primary processes of precipitation can be identified with this approach.*
3) *A multi-compartment flow sheet tool can be set up combining experimental findings and the STAR NM data to predict particle sizes.*

5.2 State-of-the-art – multicompartment approaches

Literature has provided several ideas in the last few decades which consider the abstraction of the complex control volume into several interconnected sub-zones. Gösele and Kind [109] and Van Leeuwen et al. [110] (idea 1) propose, considering phenomenological factors, a three zonal approach as a compensatory flowsheet for a continuously operated precipitation reactor.

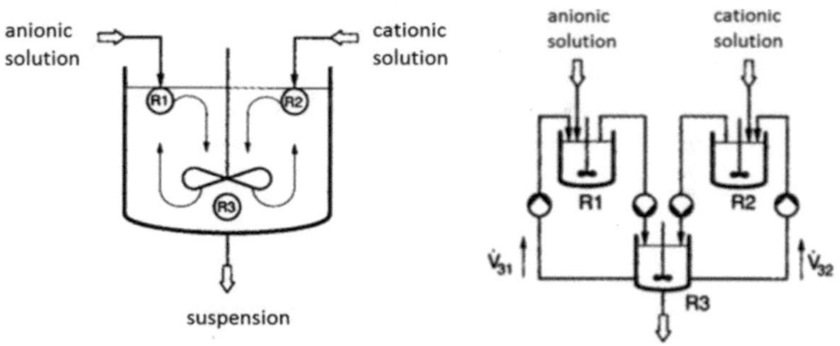

Figure 5.2: Compartment idea proposed by Gösele and Kind [109] to abstract a continuous precipitation reactor into three sub-compartments (idea 1).

Thereby, two reaction zones, R1 and R2, are connected to a third vessel, R3, representing the bulk volume (see Figure 5.2), where the product suspension is removed.

Zauner and Jones [107] (idea 2) introduced a similar flow sheet structure with their "segregated feed model (SFM)" (Figure 5.3, left), wherein the two reaction plumes, f_1 and f_2, exchange mass between each other and the bulk. One basic assumption is the instantaneously perfect mixture of the single compartments. It is important to note that only nucleation occurs in the zones f_1 and f_2; growth and aggregation are considered in the bulk phase.

Alexopoulos et al. [111] (idea 3, see Figure 5.3, right) propose a further simplified model wherein the complex hydromechanics are separated into a smaller impeller zone of high turbulent intensity and a circulation zone, where the energy dissipation ε_{circ} is significantly lower.

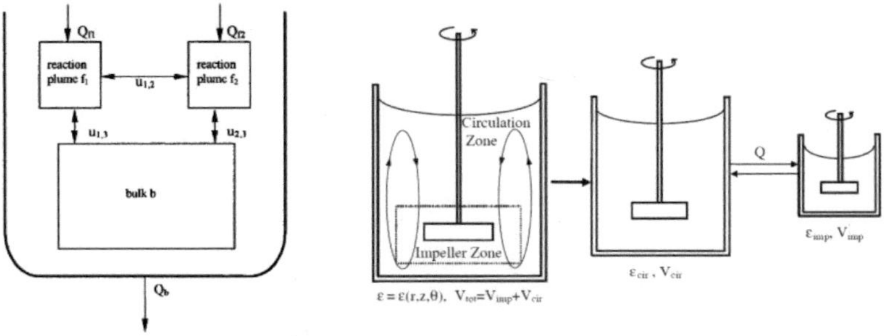

Figure 5.3: Left: Compartment structure presented by Zauner and Jones [107], (idea 2); right: Proposal of Alexopoulos [111] to abstract a vessel into an impeller zone and a circulation zone (idea 3).

More complex approaches, as exemplarily shown from Bourne and Yu [112] and Kougoulos et al. [113], see Figure 5.4 (idea 4), try to increase the accuracy of their balancing by increasing the number of compartments which are characterized by quasi-similar flow conditions.

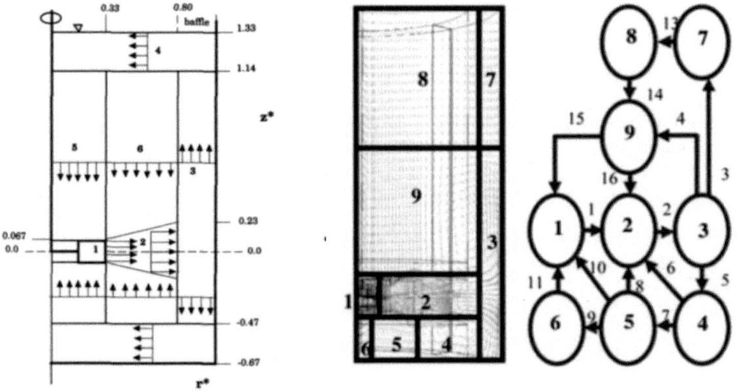

Figure 5.4: Left: Multizone model of [112], wherein the ST reactor is classified into different regions, right: Multizone model of Kougoulos et al. [113] (idea 4).

Kulikov et al. [114] established an adaptive, CFD-coupled compartment methodology which automatically reset the zonal boundaries in the transient flow field for every time step regarded (idea 5). Thereby, the compartment borders are chosen with a certain cutoff value, so that the zones created possess almost similar energy dissipation rates.

Although a group of approaches exist in literature, the adequate prediction of particle sizes for fast precipitation reactions in ST reactors is still a challenging open question. Idea 1, for example, is mainly an intellectual game. Van Leeuwen et al. [110] could show that the principal idea of a multizonal approach may be an important tool to gain an understanding of complex processes. However, first modeling results differ significantly to experiments. Idea 2 [107] could gain impressive results for slow precipitating systems (e.g. supersaturation $S_{a,CaCO3,wateq} = 12.3$ used), where the feed is introduced with a certain delaying mesomixing time and, due to comparatively slow processes, the complete vessel, except for the small unmixed feed plumes, is balanced in terms of mass, components and the population density.

Idea 3, in addition to the multizone approaches of idea 4, defines its compartment size by an energy dissipation cutoff. An average value or a radially changing progress is calculated inside the compartment. The complexity of calculations increases strongly with the number of compartments. Such allocations need extensive correlations of flow rates from one to the other compartment and of the specific energy dissipations inside a zone, which makes the model highly elaborate and relatively stiff for a flexibility requiring flow sheet simulations.

Kulikov [114] made probably one of the approaches which was most advanced within this topic. However, the adaptive compartment creation within CFD is stretched to its limits, since the multiscale character of processes (e.g. semi-batch process in the range of ~min) cannot be covered.

5.3 Simulation strategy

This work tries to tackle existing deficiencies by incorporating promising partial aspects from literature combined with the advancement of our own approach. In the following, two new theoretical approach variations are presented which are postulated to be adequate equivalent circuits for the flowsheet simulation of technical ST semi-batch processes ($V_{\mathrm{ST}} > 10\,l$, see later in 5.4.1). Only the interaction of two interconnected process zones is generally considered for the semi-batch operations investigated. Thereby, both variations involve a reaction zone, R_1, and a holdup zone, R_2. Extensions to a continuous process with a second feed may be implemented easily by adding a second interconnected reaction zone. This work focuses on a feed position close to the stirrer, where the highest energy dissipation rate for the Rushton turbine used exists (see Figure 5.1). This position is supposed to be the most representative in an ST reactor, since the directional radial flow rate promoted by the impeller is relatively easy to describe in terms of mass flow and velocity. The basic important assumptions that are made for the flowsheet model presented are pointed out in the following.

- Primary, supersaturation-dominated processes (nucleation, molecular and aggregative growth) are zonally separated from secondary processes (agglomeration, aging). Since the time scales of the particular mechanisms differ strongly (primary processes ~ms-s, secondary processes ~min-h), this assumption represents a scale separation of mechanisms. Primary processes, by definition, occur in the reaction zone R_1, secondary processes occurring globally may take place in the holdup zone R_2 (bulk). This means that the size of the reaction zones is not specified from the beginning. Precisely, the size takes the dimensions that are required to provide for the residence time needed till the supersaturation is depleted via nucleation and growth. (see, subsequently, the population balancing zones in Figure 5.26 and Figure 5.29)
- The holdup zone is considered to be perfectly mixed. Particles from R_1 are accumulated during the semi-batch process. A specific (particle loaded) mass flow $\dot{M}_{\mathrm{ent}}$ is continuously re-entrained into the reaction zone. A population balance solver that accounts for secondary mechanisms can be coupled easily to the holdup balancing, but will not be regarded within this work.

- The reactant concentrations for the $BaSO_4$ precipitation are chosen to be $S_{a,\mathrm{nom}} = 1000$ and $\mathrm{R_c} = 5$, which allow for neglecting secondary mechanisms affecting the particle collective, to reduce the interplay of the various possible mechanisms (see section 2.4, point 6). Moreover, the reactant $BaCl_2$ is filled initially into the tank, whereas the second reactant, Na_2SO_4, is fed into the reaction zone externally along the process time. This leads to a strong excess of Ba^{2+} ions, which electrostatically stabilize the primary particles (see section 2.4, point 2). A reduction of the mechanisms occurring that veil the analysis of the compartment interplay is beneficial for a proper process understanding.

5.4 Experimental setup

The laboratory plant which is used to investigate either the behavior of the precipitation ST reactor R_2 solely, or the two-compartment approach including R_1 is shown in Figure 5.5.

The basic units involved are the feed assembly, the reaction zone R_1 and the holdup zone R_2 which is also used for reference ST investigations. One concept realizing the reaction zone R_1 leans towards the suggestion of Gösele and Kind [109], using a ST reactor for the approximation. A second concept, which is generically more akin to the real mixing situation, close to the stirrer, is a jet in crossflow (JiC) arrangement. A detailed explanation of a reasonable design of these apparatuses, R_1 and R_2, will be given in the following sections.

The feeding strategy has been carefully implemented within the process. Undesirable initial contact of the reactant solutions, leading to undefined nucleation, may ruin the run. Therefore, all relevant feed pipes and ducts are flushed with compressed air initially. After the adjustment of the feed-flowrate desired by a gear pump (see flow adjustment loop in Figure 5.5), the valve, V01, is switched on and the feeding routine starts. A peristaltic pump (Bredel, APEX 20H) provides for the circulation flow of the suspension in the compartment arrangement.

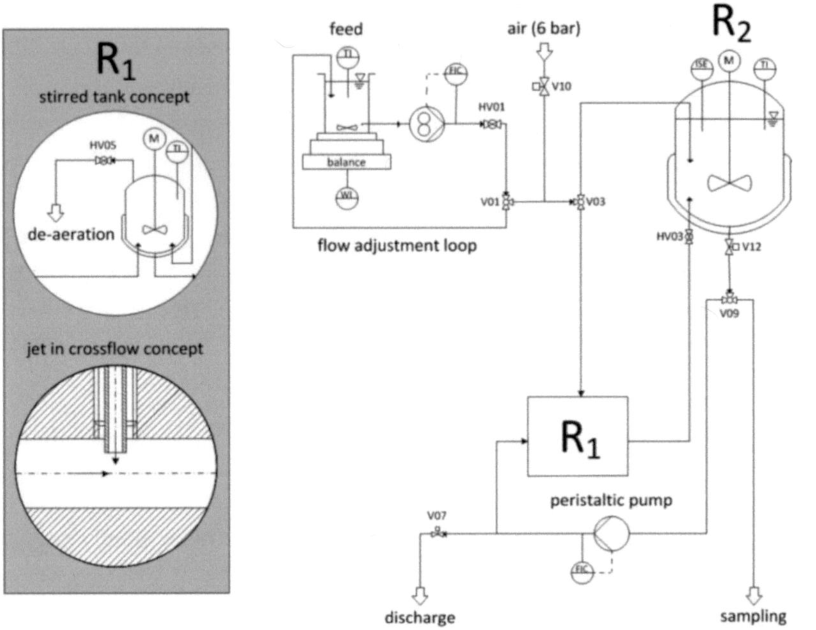

Figure 5.5: Flow sheet of the laboratory plant used for the experimental multicompartment investigations. Left: Reaction zone R_1-concepts investigated.

5.4.1 Design of the holdup vessel R_2

The main vessel (volume: $11\ l$), representing the reference ST, and the holdup zone R2 (see Figure 5.5) is constructed according to DIN 28131 [108] including four baffles. A six-bladed Rushton turbine, transporting the fluid in a radial direction, was used. The Rushton turbine represents a stirrer type used commonly for precipitation purposes [81]. It ensures a good dispersion of nano- and microparticles and, moreover, allows for the maximal feasible energy dissipation in stirred vessels which are necessary for fast mixing. Consonant with the literature [81], the mean energy dissipation in an ST can be calculated consonant with eq. 5.1. The validity of this basic equation, which is taken for reference, is investigated numerically in section 5.5.

$$\overline{\varepsilon} = \frac{Ne \cdot n^3 \cdot d_2^5}{V_{\text{Tank}}} = \frac{4 \cdot Ne \cdot n^3 \cdot d_{2,R_2}^5}{\pi \cdot {D_{R_2}}^2 \cdot H_{R_2}} \qquad 5.1$$

Regarding the model assumptions made in section 5.3, a spatial and temporal decoupling of primary and secondary processes is desired. Specifying the vessel size for the technical design, it has been considered that either the depletion of supersaturation (τ_{depl}) or the mixing process (τ_{mix}) has to be terminated within one circulation (τ_{circ}, eq. 5.2), depending on the limiting slower step. A conservative threshold value taken from the CIJM-STAR NM (section 3.4.3) is chosen for the acquisition of a minimal possible ST volume V_{Tank} that fulfills the requirement of such a separated reaction and holdup zone. For low energy dissipation which is a reasonable value for STs [56] (see Figure 3.41, $\overline{\varepsilon} = 0.1\, W/kg \;\; \rightarrow \tau_{\text{mix}} \approx 0.3\, s$) or slow supersaturation depletion (see Figure 3.42, for $S_{a,\text{min}} = 120 \rightarrow \tau_{\text{depl}} \approx 0.3\, s$),

$$\tau_{\text{circ}} = \frac{V_{\text{tank}}}{\dot{V}_{\text{circ}}} = \frac{\pi \cdot {D_{R_2}}^3}{4 \cdot f_{\text{circ}} \cdot n \cdot (0{,}33 \cdot D_{R_2})^3} \qquad 5.2$$

A mixing time of 0.15 s in the region close to the stirrer, which coincides well with the values chosen, is prognosticated for an ST with $\overline{\varepsilon} = 0.15$ W/kg and a Rushton turbine, as shown in Figure 5.6, left, [115]. Moreover, a temporal safety factor of $f_t = 5$ is added (eq. 5.3).

$$\tau_{\text{depl},f} = f_t \cdot \max(\tau_{\text{depl}}, \tau_{\text{mix}}) \qquad 5.3$$

Thus, a minimal tank diameter D_{R_2} is estimated that has a circulation time of $\tau_{\text{circ}} = 1.5\, s$. Combining eq. 5.1 with 5.2, one gets this minimal required diameter according to eq. 5.4 (visualized in Figure 5.6).

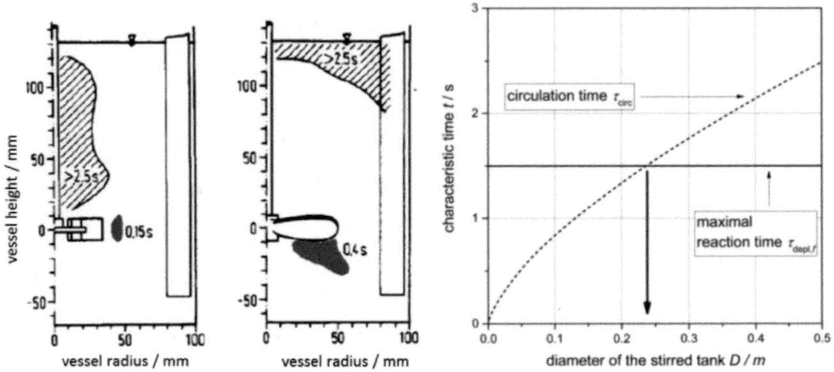

Figure 5.6: Left: Mixing times prognosticated by Geisler et al. [115] at $\overline{\varepsilon} = 0.15\,W/kg$, $\mathrm{Sc} = 1000$ and $\nu_{\mathrm{kin}} = 10^{-6}\ m^2/s$; a) Rushton turbine, b) propeller stirrer. Right: Characteristic times calculated depending on the tank diameter.

$$D_{R_2} = \frac{4}{\pi} \cdot 0.33^2 \cdot \sqrt{(f_{\mathrm{circ}} \cdot \tau_{\mathrm{circ}})^3 \cdot \frac{\overline{\varepsilon}}{Ne}} = 0.234\,m \qquad 5.4$$

Assumptions that are made in eq. 5.4:

- $d_{2,R_2} = 0.33 \cdot D_{R_2}$ [108]
- $H_{R_2} = D_{R_2}$ [108]
- $Ne = 4$ [116]
- $f_{\mathrm{circ}} = 1.5$ [117]
- conservative value (≅ fast circulation) for $\overline{\varepsilon} = 1$ W/kg [115]

Figure 5.7 shows the ST reactor constructed with baffles and the Rushton turbine. The shaft bearing is arranged in the top. Specific geometry parameters can be extracted from Table 5.1.

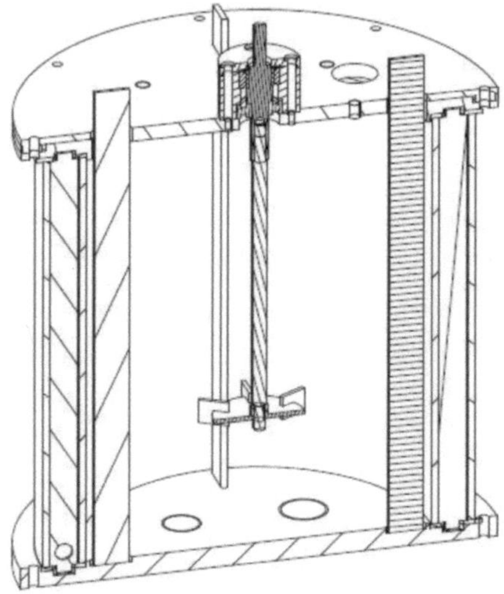

Figure 5.7: $11\ l$ *ST reactor used for the experimental investigations.*

Table 5.1: Geometrical dimensions of the bulk vessel chosen, consonant with DIN 28131 [108].

Specifications	Geometry parameters	Dimensions of the bulk vessel
Reference dimensions	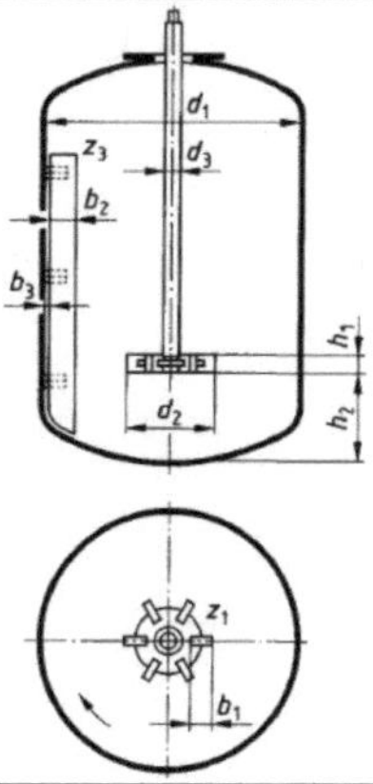	$D = d_1 = 240\ mm$
$d_2 = (0.3 - 0.4) \cdot d_1$		$d_2 = 84\ mm$
$h_1 = 0.2 \cdot d_2$		$h_1 = 16.8\ mm$
$h_2 = d_2$		$h_2 = 84\ mm$
$b_1 = 0.25 \cdot d_2$		$b_1 = 21\ mm$
$b_2 = 0.1 \cdot d_1$		$b_2 = 24\ mm$
$b_3 = 0.02 \cdot d_1$		$b_3 = 4.8\ mm$
$z_1 \geq 6$		$z_1 = 6$
$z_2 \geq 2$		$z_2 = 4$

5.4.2 Design of a stirred-tank reaction compartment – experimental concept A

The experimental reaction compartment is designed either as an ST reactor (concept A), as proposed by Gösele and Kind [109] (see section 5.2), or as a JiC arrangement (concept B, section 5.4.2). For concept A, the ST reactor, the design assumptions made are explained in the following.

- Since a reaction time $\tau_{\mathrm{depl},f}$ of $1.5\ s$ should be guaranteed (section 5.4.1), the dimensions of the reaction ST are chosen according to eq. 5.5.

$$\tau_{\mathrm{depl},f} = \frac{V_{R_1}}{\dot{V}_{R_1}} = \frac{\pi \cdot D_{R_1}^3}{4 \cdot (\dot{V}_{\mathrm{ent}} + \dot{V}_{\mathrm{feed}})} \overset{!}{=} 1.5\ \mathrm{s} \qquad 5.5$$

  following [108] with $H_{R_1} = D_{R_1}$. One crucial point is the quantity of $\dot{V}_{\mathrm{ent}}$ (eq. 5.15) which must be estimated with CFD methods (STAR NM), since it is not accessible experimentally.

$$\dot{V}_{\mathrm{ent}} = f_{\mathrm{ent}} \cdot \dot{V}_{\mathrm{prim}} = f_{\mathrm{ent}} \cdot f_{\mathrm{prim}} \cdot n_{R2} \cdot d_{2,R_2}^3 \qquad 5.6$$

  Considering the findings from section 5.5.2.1, in a first assumption, f_{ent} is chosen to be constant to 0.05 (see detailed explanation in section 5.5), and f_{prim} was chosen to be 0.8 [117], which leads, with eq. 5.5 and 5.6, to a reaction tank diameter of $D_{R_1} = 0.06\ m$.

- The local energy dissipation $\varepsilon_{\mathrm{feed}}'$ at the feed position is extracted from the CFD, which conforms to the literature [56] (see Figure 5.17), and is estimated to be the mean energy dissipation $\bar{\varepsilon}_{\mathrm{R_1}}$ in the reaction compartment R_1 (details are given in section 5.5). A value of $\phi_\varepsilon = 6$ is selected (eq. 5.7) for the feed position specified in Table 5.9.

$$\bar{\varepsilon}_{R_1} = \varepsilon_{\mathrm{feed}}' = \phi_\varepsilon \cdot \bar{\varepsilon}_{R_2} \qquad 5.7$$

  Considering the diameters D_{R_1} ($= 0.06\ m$), D_{R_2} (Table 5.1) and eq. 5.1, one gets eq. 5.8, which describes the ratio between the two rotational speeds in the two compartments.

$$n_{R_1} = \sqrt[3]{96} \cdot n_{R_2} \qquad 5.8$$

Table 5.2: Geometrical dimensions of the ST reaction compartment $R_{1,\mathrm{CSTR}}$ chosen according to DIN 28131

Geometrical parameters [108]	Size or number
$D_{R_1} = d_1$	$60\ mm$
$d_2 = 0.35 \cdot d_1$	$21\ mm$
$h_1 = 0.2 \cdot d_2$	$4.2\ mm$
$h_2 = d_2$	$21\ mm$
$b_1 = 0.25 \cdot d_2$	$5.5\ mm$
z_1	6

Table 5.2 shows the geometrical dimensions of the CSTR reaction compartment. The complete apparatus realized is shown in the appendix in Figure 8.19, left. The volume ratio between the two STs, R_2 and R_1, is $V_{R_2}/V_{R_1} = 70$.

5.4.3 Design of a jet in crossflow reaction compartment – experimental concept B

Concept A assumes a uniform energy dissipation rate in the reaction compartment. This simplified assumption, which is also made for the mixing zone in CIJMs, may be questioned critically for the far less concentrated mixing situation in ST reactors. Therefore, a JiC structure is chosen for the reaction compartment R_1 to create a generically more similar mixing environment (see Figure 5.8).

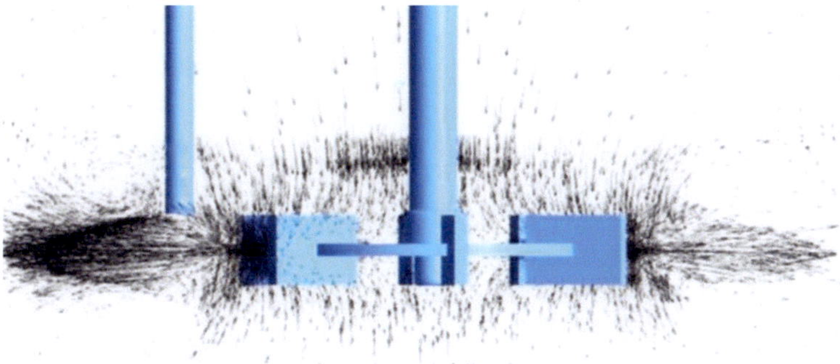

Figure 5.8: Feed situation investigated in the reference vessel R_2; a snapshot from CFD.

The basic assumptions chosen are:

- The mean crossflow velocity in the JiC compartment is chosen similar to the local velocity u_{feed,R_2} of the global flow field in the ST at the feed position. This local velocity is estimated consonant with eq. 5.9, calculating the mean velocity of the stirrer primary flow, passing a shell surface of a cylinder (with the height h_{1,R_2}) at the feed position. Equation 5.10 represents the flow rate through the crossflow compartment.

$$u_{\text{feed},R_2} = \dot{V}_{\text{prim}} / (2 \cdot \pi \cdot r_{\text{feed},R_2} \cdot h_{1,R_2}) \qquad 5.9$$

$$\dot{V}_{\text{ent}} = \frac{\pi}{4} \cdot d_{i,R_1}^2 \cdot u_{\text{feed},R_2} = f_{\text{ent}} \cdot \dot{V}_{\text{prim}} \qquad 5.10$$

Combining eq. 5.9 and 5.10, this leads to a pipe diameter according to eq. 5.11, which depends only on the stirrer height h_{1,R_2} and the radial position of the stirrer r_{feed,R_2}. one single pipe of diameter d_{i,R_1} is needed for a specified feed position, independent of the stirrer speed.

$$d_{i,R_1} = \sqrt{8 \cdot f_{\text{ent}} \cdot h_{1,R_2} \cdot r_{\text{feed},R_2}} \qquad 5.11$$

- The parabolic radial velocity profile generated by the stirrer (Figure 5.9, left) is, in a first assumption, akin to the not fully developed Hagen-Poiseuille profile which exists in the crossflow.

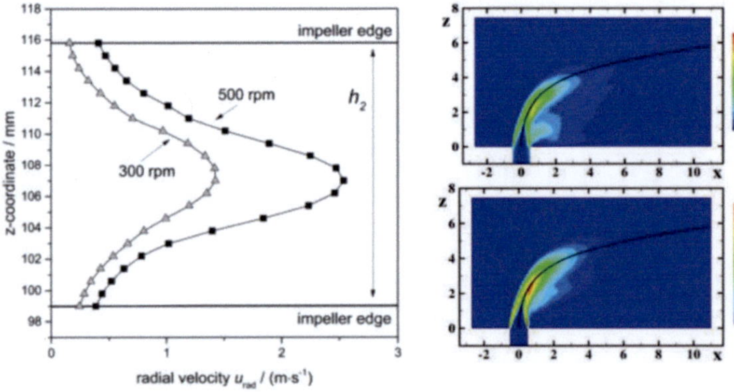

Figure 5.9: Left: Flow profile of the stirrer extracted from CFD (see section 5.5), right: Turbulent kinetic energy and specific rate of dissipation ω in the crossflow situation, cited from Denev et al. [118].

- The main turbulence in the system in terms of k and ε is induced by the JiC arrangement. Denev et al. [118] showed that high turbulence is created mainly at the jet surface where the crossflow hits the jet stream (see Figure 5.9).
- The length of the compartment can be estimated by the demand of 1.5 s "reaction time" $\tau_{\mathrm{depl},f}$ (see chapter 5.4.1). $\dot{V}_{\mathrm{ent}}$ is calculated consonant with eq. 5.15, with $f_{\mathrm{ent}} = 0.05$, resembling concept 1.

$$\tau_{\mathrm{depl},f} = \frac{\pi \cdot d_{i,\mathrm{crossflow}}^2 \cdot L_{\mathrm{crossflow}}}{4 \cdot (\dot{V}_{\mathrm{ent}} + \dot{V}_{\mathrm{feed}})} \stackrel{!}{=} 1.5\ \mathrm{s} \tag{5.12}$$

- The feed pipe is immersed with $h_{\mathrm{feed}} = 3\ mm$ into the flow field to avoid wall interactions of the mixing jet flow (e.g. horseshoe vortices and wake structures [119]) which are not present in ST reactors.

Considering the assumptions, geometrical dimensions are chosen as shown in Table 5.3. The geometry constructed is pictured in Figure 5.10.

Table 5.3: Geometrical dimensions of the jet in crossflow (JiC) reaction compartment $R_{1,JiC}$ chosen following DIN 28131

Geometrical parameters	Size or number
d_{i,R_1}	$19\ mm$
L_{R_1}	$659\ mm$
$d_{i,\mathrm{feed}}$	$4\ mm$
h_{feed} (= immersion depth of the feed pipe)	$3\ mm$

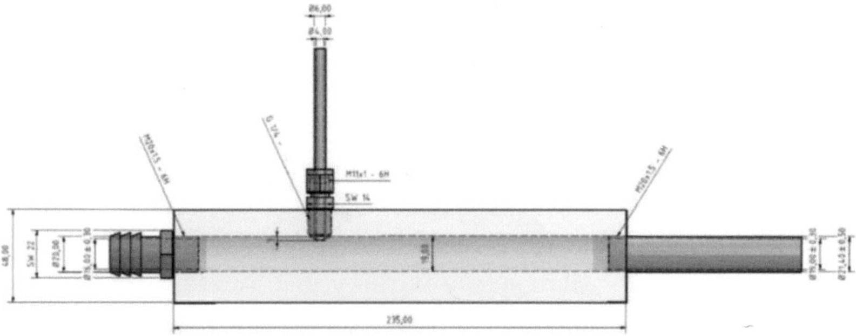

Figure 5.10: Experimental setup of the JiC section, whereby the mixing-zone is made of PMMA for visual access (see also appendix Figure 8.19, right).

5.5 Numerical section – estimation of process-dominating parameters by STAR NM

This section provides insight into the dominating flow mechanics in ST reactors. Similar to the investigations carried out for the CIJMs, CFD provides a precious way to gain insight into inaccessible processes and to "numerically measure" quantities that are not measurable experimentally. Values such as the entrained mass flow $\dot{M}_{\mathrm{ent}}$, which is circulated by the peristaltic pump, are unknown for the design of precipitation experiments and for the flow sheet simulation. Missing values and critical aspects are introduced in this section and compared, when possible, to literature and discussed.

5.5.1 Numerical setup

Numerical investigations are executed using the Euler-Euler two-phase model. Thereby, three phases are selected (primary phase: air, secondary phase 1: $BaCl_2$-water, secondary phase 2: Na_2SO_4-water). Further simulation details are given in Table 5.4.

Table 5.4: Simulation features chosen for the ST reactor simulation

Features	Approach selected
mode of calculation	transient
simulation approach	Euler-Euler model, species transport
turbulence model	k-ε RNG model (swirl dominated flow)
pressure-velocity coupling	Phase-coupled simple
discretization	Higher-order discretization (third-order MUSCL, bounded central differencing)

The numerical 3D grid used is shown in Figure 5.11. The impeller is embedded into a bounding cylinder and additionally covered with eight prism layers on the overall impeller surface to resolve the hydrodynamic boundary layer to account for the rotating parts in the stirred vessel. A second surrounding static cylinder with the same refined grid parameters is placed to ensure a proper scalar transport over the sliding mesh interface. Regions of high gradients and zones of enhanced interest in the radial section of the vessel around the impeller are further refined.

Table 5.5: Overview of the specific mesh quality parameters. On the one hand, the recommendation by Ansys Fluent [120] and, on the other hand, the real mesh parameters are shown.

Criteria	*Recommendations*	*Max/min values*
orthogonal quality	$min\ 0.01$	0.1038
skewness	$max\ 0.95$	0.732
aspect ratio	$max\ 10$	6.01

The feed inlet zone is especially resolved by adding two spherical refinements. Table 5.5 gives an overview of the specific mesh quality parameters.

Although it has been shown (in Figure 3.17) for the CIJM that the URANS k-ε model may overestimate the mixing velocity, in this context, the k-ε Re normalization group (RNG) model is used as a literature-recommended turbulence model suitable for ST reactors [121, 122]. Thereby, Karimi et al. [122] could show that the k-ε RNG model for ST reactors – in terms of turbulence and velocity distribution – delivers results that fit well to experimental data.

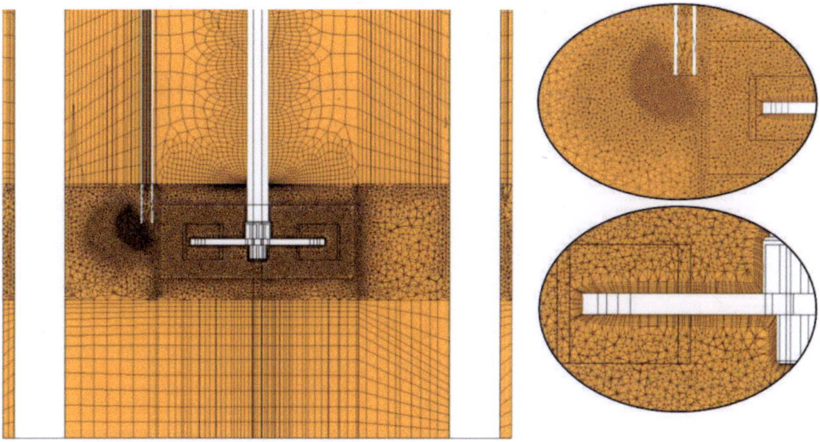

Figure 5.11: Right: Sectional view of the mesh used for the numerical investigations with 3,267,500 cells and 984,200 nodes. Right: Highlighted features, top: Two levels of spherical refinements in the feeding zone, bottom: Prism layers on the impeller surface to resolve the hydrodynamic boundary layer.

This approach is chosen as a compromise, on the one hand, due to turbulence model limitations arising from the two-fluid approach and, on the other hand, due to a reasonable computational effort. When interpreting the mixing "measurements" in section 5.5.2, the reader should keep this potential over-prediction in mind. The k-ε RNG model and the sliding mesh procedure are explained briefly in the appendix.

5.5.2 Numerical results

5.5.2.1 Estimation of dominating circulation rates

Circulation in ST reactors is classified into dominant flow phenomena. In the literature, the flow field generated by a Rushton turbine is categorized into a primary flow $\dot{V}_{prim}$ (eq. 5.13) promoted directly by the stirrer of diameter d_2 at the rotational speed n which passes a lateral area of the stirrer height h_1.

$$\dot{V}_{\mathrm{prim}} = f_{\mathrm{prim}} \cdot n \cdot d_2^3 \qquad 5.13$$

The pumping number f_{prim} describes the fraction of the totally pumped stream which is driven to the vessel wall, whereas some fluid gets lost by recirculation at the boundary (see Figure 5.8).

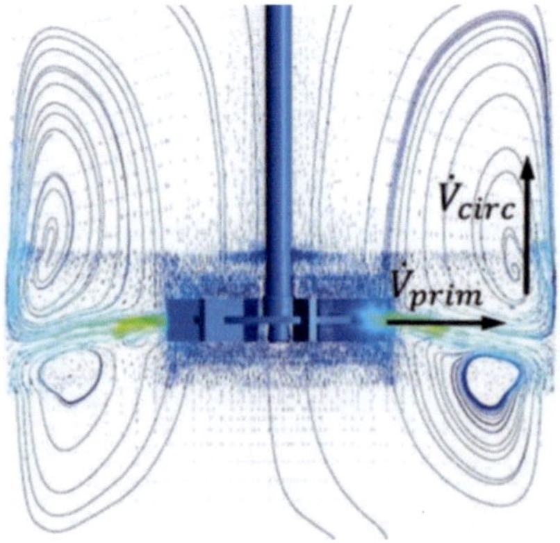

Table 5.6: Overview of values for f_{prim}, cited from Baldyga et al. [117]

Author	f_{prim}
Sachs and Rushton (1954)	0.47
Norwood and Metaner (1960)	0.85
Cooper and Woolf (1968)	0.73 – 0.89
Desouza and Pike (1972)	0.86
Reed et al. (1977)	0.65
Helmstaedtter (1981)	0.85
Revill (1982)	0.75 ± 0.15
Yanneskis et al. (1987)	0.785
Costes and Couderc (1988)	0.73
Wu and Patterson (1989)	0.86
Ranade and Joshi (1990)	0.75

Figure 5.12: Left: Classification of predominant flows in an ST reactor with a Rushton turbine; vector plot with streamlines of the ST R_2.

Table 5.6 shows specific pumping numbers cited from Baldyga et al. [117] that describe the impeller discharge. Our own numerical investigations in the $11\ l$ vessel confirm values in the range of 0.75 ± 0.25 (see Table 5.7) for the turbulent regime. A significantly lower discharge is measured for $50\ rpm$ (transition regime). In the following, especially for the flow sheet simulation of semi-batch processes, $f_{\mathrm{prim}} = 0.75$ will be used.

Table 5.7: Numbers f_{prim} (eq. 5.13) in the $11\ l$ ST obtained by measuring the volume flow through the planes A and B.

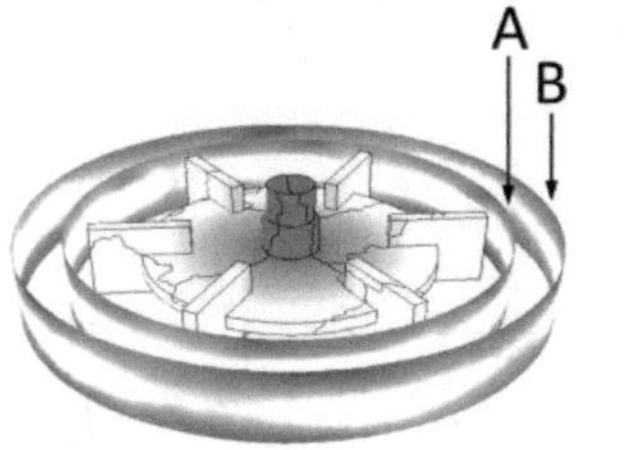

Shell plane (radial position)	*50 rpm*	*300 rpm*	*500 rpm*
A *(y = 0.392)*	*0.353*	*0.608*	*0.575*
B *(y = 0.474)*	***0.270***	***0.766***	***0.724***

It has been found that flow rates through a shell plane of h_1 have got to be measured at a certain distance from the impeller tip (see plane B) so as not be affected by recirculating fluid induced by the stirrer blades; otherwise, the pumping number is underestimated (compare values measured in shell surfaces A and B in Table 5.7). However, literature values for plane B in the turbulent regime are well reproduced. The pumping numbers to insert in eq. 5.13 for a lower impeller speed should be chosen carefully.

The second important flow rate specified in literature is the circulation stream $\dot{V}_{circ}$ in the vessel (Figure 5.12, left), which is defined similar to the primary flow according to eq. 5.14, using a circulation number f_{circ}.

$$\dot{V}_{\mathrm{circ}} = f_{\mathrm{circ}} \cdot n \cdot d_2^3 = K \cdot f_{\mathrm{prim}} \cdot \dot{V}_{\mathrm{prim}} \qquad 5.14$$

The factor $K = f_{\mathrm{circ}}/f_{\mathrm{prim}}$, which is often used in literature, is called the bulk entrainment coefficient. Table 5.8 gives an overview of specific values of K cited in literature. f_{circ} in Table 5.8 is calculated with $f_{\mathrm{prim}} = 0.75$. Numerical investigations indicate that a circulation flow rate with a constant f_{circ} may be a simplified assumption.

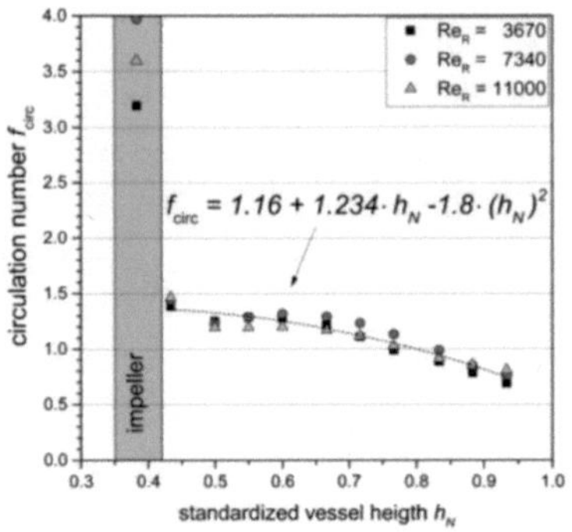

Table 5.8: Bulk entrainment coefficients and circulation numbers cited from Baldyga et al. [117]

Author	K	f_{circ}
Holmes et al. (1964)	*1.9*	*1.425*
Cooper and Woolf (1968)	*1.8*	*1.35*
Rao and Brodkey (1972)	*1.5*	*1.125*
Günkel and Weber (1975)	*1.8-1.9*	*1.35-1.425*
Wu and Patterson (1989)	*1.7*	*1.275*
Ranade and Joshi (1990)	*1.6-2.0*	*1.2-1.5*

Figure 5.13: Left: Circulation numbers measured within CFD over the standardized heights h_N and for three different Re$_R$ numbers. A simple polynomial fit $f_{\mathrm{circ}} = f(h_{\mathrm{rel}})$ in the validity range of $(0.42 \leq h_N < 1)$ and for all $\mathrm{Re_R}$ numbers investigated is proposed.

Moreover, a decreasing value for the circulation number is expected with a decreasing distance to the liquid surface, since fluid is re-entrained towards the impeller. However, circulation numbers measured up to $0.7 \cdot h_N$ $(h_N = h/H)$ fit well to values listed in literature (see Table 5.8, [117])

The fluid fraction of $\dot{V}_{\mathrm{prim}}$ that is entrained into the feed stream and which is involved dominantly in mixing (see Figure 5.12, right) is expressed in this work according to eq. 5.15. It is important to mention that the factor f_{ent} is not a fixed value and, moreover, is dependent on the time the supersaturation depletion takes and, thus, on the duration fluid is entrained into the reaction zone.

$$\dot{V}_{\mathrm{ent}} = f_{\mathrm{ent}}\big(\tau_{\mathrm{depl},f}\big) \cdot \dot{V}_{\mathrm{prim}} \quad 5.15$$

This factor which plays an exceptional role within the flow sheet model presented in this chapter is not accessible by experiment. Therefore, STAR NM is used to estimate this factor. Figure 5.14 shows an isosurface of feed and entrained volume flow at $n = 500\ rpm$ and $\dot{V}_{\mathrm{feed}} = 400\ ml/min$. The counter-rotating vortex pair [119], which is developed and which plays an important role in literature for the fluid entrainment in such a JiC situation is clearly recognizable. As long as this vortex pair exists $\approx y_N \leq 0.73$, entrainment is strongly promoted. Afterwards, the rate of fluid involved in mixing decreases (Figure 5.14, right). Moreover, higher feed rates entraining more liquid is observed, which is easy to envision. We determine by the CFD "measurements" the proportionality that twice the feed rate may entrain twice the mass of fluid. The initial quasi-linear slope m_0 in Figure 5.14, right, almost identical at differing rotational speeds observed, can be empirically characterized with the simple linear relation (eq. 5.16).

$$\begin{aligned}\dot{V}_{\mathrm{ent}} /(m^3 s^{-1}) &= m_0 \cdot \dot{V}_{\mathrm{prim}} \cdot t \\ &= 2.594 \cdot 10^5 \cdot \dot{V}_{\mathrm{feed}} \cdot \dot{V}_{\mathrm{prim}} \cdot t\end{aligned} \qquad 5.16$$

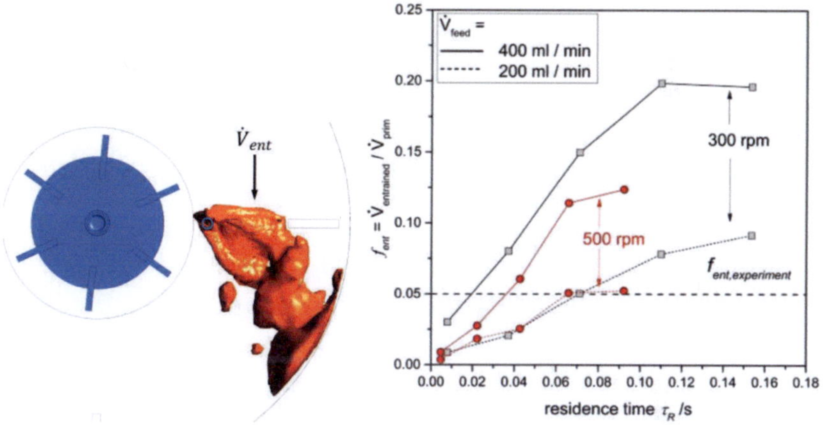

Figure 5.14: Left: Screenshot from CFD ($n = 500\ rpm$ and $\dot{V}_{\mathrm{feed}} = 400\ ml/min$) indicating that only a fractional amount of the overall pumping flow $\dot{V}_{\mathrm{prim}}$ is entrained into the reaction zone. Right: Measurement of the entrained fraction f_{ent} of the pumping flow. Additionally, the experimentally fixed value $f_{\mathrm{ent}} = 0.05$ (see section 5.6) is added.

5.5.2.2 Residence times

Following the STAR NM approach (section 3.4.3, step 2), residence time can be measured by inserting tracer particles into the system (see Figure 5.15) and by measuring the time particles need to form the entrance plane to a specific screening plane, e.g. the ones used in Figure 5.18. In this section, in a first assumption, only a simplified approach, which may be refined for future works, is chosen due to computational costs.

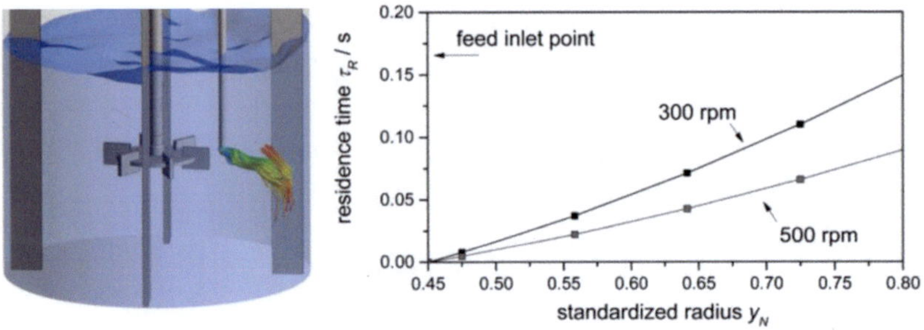

Figure 5.15: Left: Streamlines of fluid fed into the ST. Right: Simplified calculation of residence times with eq. 5.18.

$$u_i = \frac{\dot{V}_{\mathrm{prim}}}{\pi \cdot y_{N,i} \cdot D \cdot h_1} \qquad 5.17$$

$$\tau_{R,R_2} = \sum_{i=0}^{k} \left((y_{N,i+1} - y_{N,i}) \cdot D_{R_2} / \left(\frac{u_i + u_{i+1}}{2} \right) \right). \qquad 5.18$$

Residence times are calculated by eq. 5.17 and 5.18, taking into account only the radial decrease of velocity due to the enlargement of the shell surface of the normalized radius y_N and the height h_1, which is passed through by the primary flow. The additional feed stream may increase the velocity locally, but is not considered here. f_{prim} is chosen to be 0.75 (see section 5.5.2.1).

5.5.2.3 Mean and local energy dissipations

Mean energy dissipation rates are measured within CFD for different relative liquid levels ($h_{R_2}^L = H_{R_2}^L / D_{R_2}$) and rotational numbers (see Figure 5.16). Comparison is made to the values predicted by eq. 5.1, which is commonly used in literature [81, 123].

It can be stated that $\bar{\varepsilon}$ measured in CFD differs from mean values calculated with eq. 5.1. Deviations for $h_{R_2}^L = 0.5$ are expected, since the impeller is not fully submerged into the liquid (see also Figure 5.20) and a correct power input is not fully given. The effect strengthens for higher rotational numbers. One aspect in eq. 5.1 that should be discussed critically is the point that internals, such as baffles or the feed pipe, influencing the turbulence significantly are not considered.

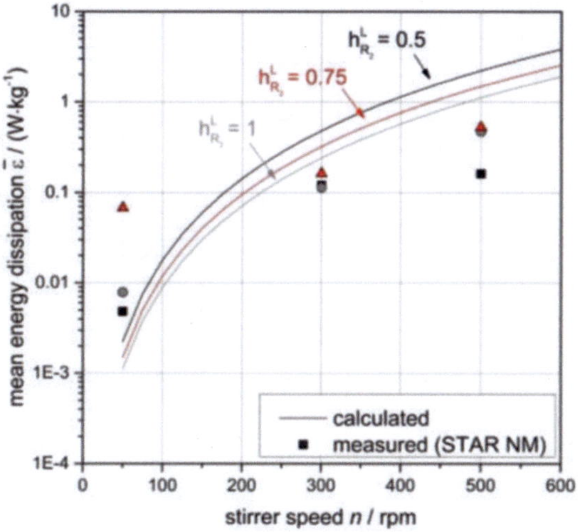

Figure 5.16: Mean energy dissipation measured in R_2 for different rotation numbers and liquid levels. The lines are calculated with eq. 5.1.

Moreover, no range of validity is given, which means that it is not defined whether this simplified approach may be valid for every flow regime.

Additionally, local relative energy dissipations $\theta = \varepsilon/\bar{\varepsilon}$, measured within CFD are compared to values predicted in literature. In detail, θ is calculated using both the local and the mean value (see dots in Figure 5.16) of ε from CFD. Laufhütte and Mersmann [56] (Figure 5.17, left) and Geisler [124] generated flow charts with isoenergetic lines of an equal θ experimentally by laser Doppler anemometry. Additionally, Baldyga and Bourne [125] present an equation (eq. 5.19) that describes the radial progress of θ at the stirrer position ($y_N = 0$, see Figure 5.17, left).

$$\theta = \frac{\varepsilon}{\bar{\varepsilon}} = \frac{1.2}{(y_N - 0.23)^2} \qquad 5.19$$

Two conclusions can be drawn regarding Figure 5.17. One the one hand, the different sources cited are in good accordance to each other. On the other hand, one can see that for the fully turbulent $\mathrm{Re_R}$, "measurements" from CFD suit literature well, although the mean values differ significantly.

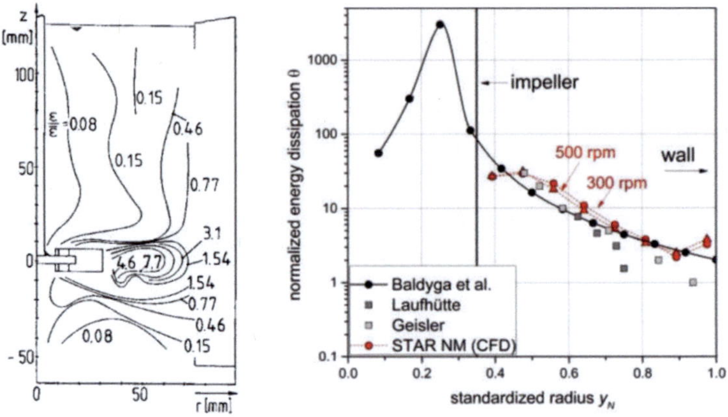

Figure 5.17: Left: Energy dissipation map from Laufhütte and Mersmann [56], at $\mathrm{Re_R} = 31\,000$, right radial values at $z = 0$ (see figure left) in the impeller zone.

Thus, eq. 5.19 by Baldyga and Bourne [125] may act as a good approximation to describe the radial energy dissipation profile for the fully turbulent regime in STs with a Rushton turbine.

Since mixing models such as the LTSA need a temporal dependency of the energy dissipation, a conversion of eq. 5.19 must be carried out. Considering eq. 5.17 and with the assumption that $r_i = r_{\mathrm{feed}} + u_i \cdot t$, the spatial progress of the energy dissipation for a certain primary flow can be expressed time-dependently by inserting eq. 5.20 into 5.19. r_{feed} represents the radial position of the feed measured from the stirrer axis.

$$y_N = \frac{r_{\mathrm{feed}} + \sqrt{\dfrac{\dot{V}_{\mathrm{prim}} \cdot t}{2\pi \cdot h_1}}}{D/2} \qquad 5.20$$

5.5.2.4 Supersaturation and free lattice ion ratio R_C

As demonstrated for the CIJM, supersaturation buildup and the free lattice ion ratio represent the key features accounting for mixing influences. Similar to the STAR NM section in chapter 3.4.3 and the ST reactor, locally mass weighed-averaged mean supersaturation values are measured and, furthermore, averaged in time (see Figure 5.19, left).

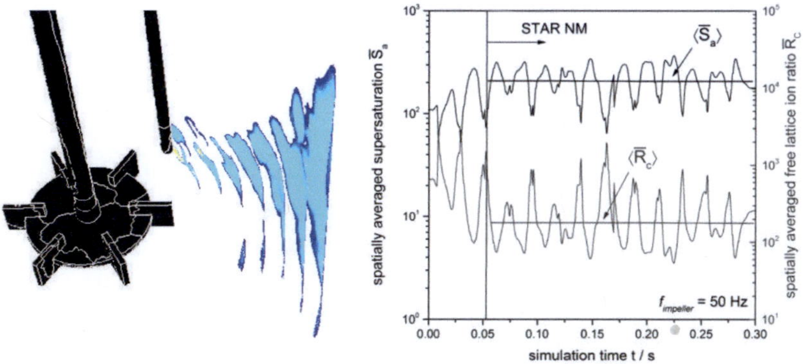

Figure 5.18: Left: Cylindrical "measurement" clip surfaces with a threshold value of $S_a = 20$ in the feed zone of the ST reactor R_2. The specific planes are assembled 1 cm from each other. Right: Exemplary spatially measured and temporally averaged supersaturation and free lattice ion ratio for $n = 500\,rpm$ and $\dot{V}_{\text{feed}} = 400$ ml/min at a 25 mm distance from the center of the feed pipe.

In order to realize a reasonable averaging only within the sectional, supersaturated zone clip surfaces are created.

A threshold value (clip value) of $S_a \geq 20$ has been chosen for the creation of these surfaces (see Figure 5.18, left). Adaptively, these specified clip surfaces are newly created for every measurement point in the transient flow field and subsequently deleted. In this way, one gets spatially averaged information for every clip plane, as demonstrated in Figure 5.18, right. As long as the Na_2SO_4 ions fed are not recycled in the circulation stream (basic assumption: SO_4^{2-} ions are consumed during precipitation within one recycle), averaging in time is performed for the quasi-stationary range. Figure 5.18, right, leads to some interesting observations. Thus, the frequency of the stirrer (in this case $50\ Hz$) plays a role clearly visible in the data measured and, moreover, in the standard deviation of the ones averaged temporally ($\rightarrow$ 14 to 15 oscillations for $0.3\ s$ plotted). Influences affecting the amplitude have not yet been investigated. The supersaturation and free lattice ion ratio are inversely arranged. This means high supersaturation leads to moderate R-values and low supersaturation causes high free lattice ion ratios. This can be understood easily regarding the precipitation diagram of the Pitzer activity coefficient (see Figure 2.4), considering that a high amount of Ba^{2+}-ions at one point in time and a low amount of Ba^{2+}-ions at another are entrained that contribute to $\langle \bar{S}_a \rangle$. Instead of an $\mathrm{R_c} = 5$, which is hypothetically adjusted for the reactants for the case of instantaneous mixing ($\mathrm{M_R} = 1$), one can see that for this semi-batch mixing situation, strongly nonstoichiometric states always shifted to values of $\mathrm{R_c} \geq 100$ can be expected. This matter involves a big challenge for the activity coefficient models used, since it is well-known that the accuracy is reduced for strong nonstoichiometric cases [7] and, moreover, diverse models differ strongly (see Figure 2.4).

Figure 5.19 shows the "measured" supersaturation profiles at two stirrer speeds and for two feed rates (200 and $400\ ml/min$) each. The dotted lines are trend lines that are extrapolated to the feed point, where no measurement plane is placed. The initial values $\langle \bar{S}_{a,\mathrm{init}} \rangle$ (indicated red) will be of further interest in the process simulation part in section 5.7. One can see that the curve differs basically from supersaturation profiles in CIJMs (see Figure 3.32). Instead of a buildup that is supposed to also exist for CSTRs straight at the feed point, but which is very fast, a diluting effect due to the entrainment of fresh bulk fluid is observed.

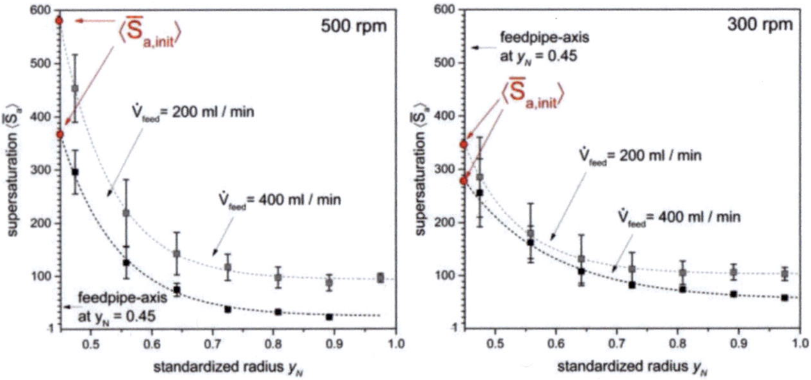

Figure 5.19: Spatially and temporally averaged supersaturation $\langle \bar{S}_a \rangle$ plotted versus the normalized radius in the ST R₂ for two different feed rates. Left: $500\ rpm$, Right 300 rpm. The dashed lines are exponentially fitted on the data and extrapolated to $y_N = 0.45$ (axis of the feed pipe, see also Table 5.1)

Thus, a slight but important mechanistic difference between reaction zones in the two apparatuses exists. The dilution of the feed occurs mainly at the first radial section, where energy dissipation is high and the counterrotating vortex phenomenon is active. Process simulation subsequently will account for this point. The maximal value reached for the specific pairing of $\dot{V}_{\mathrm{feed}}$ and n is interesting. Instead of $S_a = 1000$ and $\mathrm{R_c} = 5$, significantly lower values ($\sim S_a = 300 - 600$) are observed. This leads to the expectation of significantly larger particles due to lower nucleation rates - just as a result of the mixing point adjusted.

5.5.2.5 Semi-batch operations – critical aspects

All investigations presented in this work are carried out in semi-batch operation mode. This implies that the level of fluid initially is half of the tank height (neglecting the surplus volume of the impeller). During the process, the fluid level increases, which also implies changes in the macroscopic flow mechanics. Literature [126] points out that the flow structure below a critical height of $H^L_{\mathrm{crit}} = h_2 + 0.77 \cdot d_R$ (for $d_R/D = 0.31$) circulation and power input differs significantly. Our own CFD simulations executed for three different fluid levels in the reference ST R_2 (see Figure 5.20) support this issue, which may be important to consider for the early process stages of semi-batch precipitation.

This influence has also to be considered while regarding dominating flow rates or mean energy dissipation rates (see Figure 5.16) as one important parameter promoting mixing (see section 3.4.4).

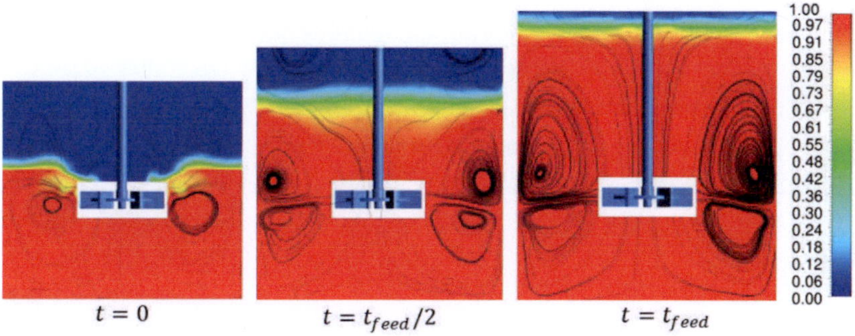

Figure 5.20: Contour plot of the liquid phase fraction with streamlines in the ST reactor at different feed times for $= 300\ rpm$.

5.6 Results – experimental investigations

Two main points are advanced in this chapter. Firstly, an experimental database of results achieved in the reference $11\ l$ ST (R_2) reactor is given. Thereby, the "material function" parameter (see chapter 1) is kept constant throughout all experiments ($BaSO_4$, $S_{a,\mathrm{nom}} = 1000$, $\mathrm{R_c} = 5$, see chapter 3.1.1) to investigate the "machine function" of the apparatus. Specific parameters depending on the apparatus influencing the precipitation are carved out in this section. Secondly, precipitation results from the two-compartment semi-batch experiments are presented and compared to the reference to verify or falsify the main hypothesis (No. 1, see chapter 5.1): A multicompartment arrangement can act as an equivalent circuit for a technical ST reactor, which this section tries to tackle.

5.6.1 Experimental reference – stirred-tank investigations

Experimental reference experiments are made in the apparatus R_2 (see Table 5.1) with a feed position close to the stirrer. The detailed specification and position of the feed pipe is given in Table 5.1.

Table 5.9: Standard adjustments investigated: Diameter and feed-position close to the stirrer chosen. Reference lengths can be extracted from Table 5.1.

$d_{i,\mathrm{feed}}$ / mm	4
$x_{\mathrm{feed}} = h_{\mathrm{feed}} / d_1$	0.4125
$y_{N,\mathrm{feed}} = r_{\mathrm{feed}} / r_1$	0.45

The influence of the feed rate and stirrer speed is investigated for the semi-batch process. Figure 5.21 with the table involved gives an overview to specify the operating conditions chosen. The $\mathrm{Re_R}$ number ($= n_{R_2} \cdot d_{2,R_2}^2 / \nu_{\mathrm{kin}}$) can be linked to the mean energy dissipation rate in the vessel (eq. 5.1), according to eq. 5.21. $\mathrm{Re_R}$ and dependent Newton numbers are taken from the VDI Heat Atlas [116].

$$\bar{\varepsilon} = \frac{Ne \cdot {Re_R}^3 \cdot \nu^3}{d_R \cdot V_{\mathrm{Tank}}} \quad 5.21$$

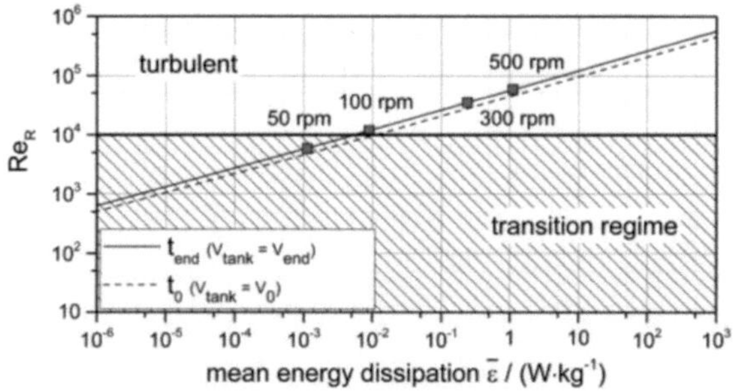

Operation points (t_{end})		
n	Re_{R}	$\bar{\varepsilon}$
rpm	-	*W/kg*
50	*5869*	*0.0011*
100	*11739*	*0.0089*
300	*35216*	*0.24*
500	*5869*	*1.11*

Figure 5.21: Operation points at different rotation numbers investigated. Thereby, the change in liquid volume is taken into account. The dashed line is the initial state and the black line indicates the end of feeding at a filled tank.

For the 11 l vessel investigated, Figure 5.21 embeds the operating conditions chosen into the framework of the flow regime and associated mean energy dissipation. The threshold value $\mathrm{Re}_{\mathrm{R}} = 10\ 000$ for the fully turbulent regime is taken from the VDI Heat Atlas [116]. As can be seen, experiments at 50 and 100 rpm are in or close to the transition regime.

Moreover, assuming the validity of eq. 5.1 for the semi-batch process, a bisectional decrease of the mean energy dissipation rate is to be expected from the start of experiments ($V_0 = 5.5\ l$) to the final state with $V_{\mathrm{end}} = 11\ l$.

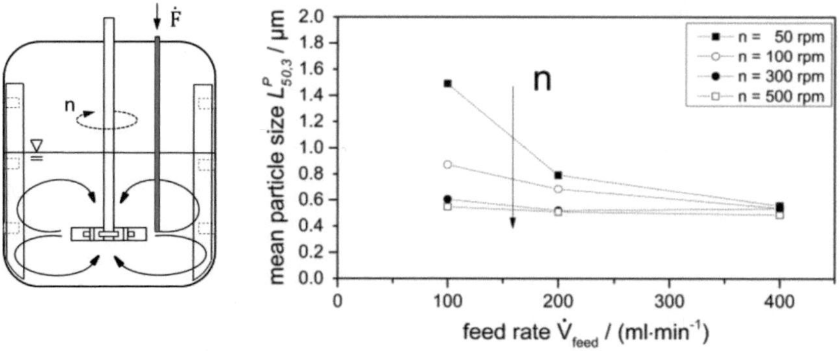

Figure 5.22: Experimentally measured mean volumetric particle sizes in semi-batch experiments (5.5 l $BaCl_2$ initially filled in the tank, 5.5 l Na_2SO_4 fed, ST, Rushton turbine, feed position according to Table 5.9), for $S_{a,\mathrm{nom}} = 1000$, $\mathrm{R_c} = 5$, at standard conditions ($\vartheta = 20\ °C, 1\ bar$).

The mean volumetric particle sizes are evaluated by static light scattering (see section 0) in a flow cell. Figure 5.22 shows reference experiments in the ST R_2 operated as described in the previous section.

Experiments indicate that an increasing feed rate and an increasing stirrer speed lead to smaller particles $L^P_{50,3}$. Experiments especially in the transition state tend to produce larger particles as was also observed for the CIJMs. Bourne et al. [127] confirm such observations made for fast reacting systems. Influences of the feed pipe diameter are varied and can be extracted from the appendix.

5.6.2 Experimental simulation – a compartment approach

In this section, the fundamental question is discussed whether a reference ST precipitation can be approximated by an equivalent circuit of interconnected compartments. Therefore, results from section 5.6.1 are reproduced within the two-compartment approach taking into account all the design assumptions made in chapter 5.6.1. Figure 5.23 shows the results for a reference experiment at $n = 300\,rpm$ and a feed rate of $100\,ml/min$ compared to concept A and concept B.

Highly congruent results can be found for the jet in crossflow concept.

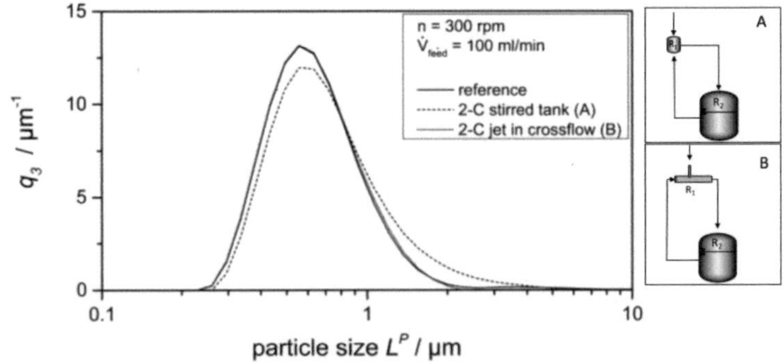

Figure 5.23: Comparison of semi-batch experiments ($S_{a,\mathrm{nom}}$ = 1000, R_c = 5) at $n = 300\,rpm$ and a feed rate of $\dot{V}_{\mathrm{feed}} = 100\,ml/min$. The PSD received in the reference experiment compared to those held by concept A (CSTR-holdup) and concept B (JiC-holdup)

Concept A using the CSTR for R_1 is also close to the reference distribution measured, but tends to deliver a broadened PSD. This observation can possibly be led back to the residence time behavior the apparatus R_1 induces in the precipitation process. Nevertheless, it can be concluded that the design assumptions made fit well. In the following, due to a generically higher similarity, further investigations focus on concept B.

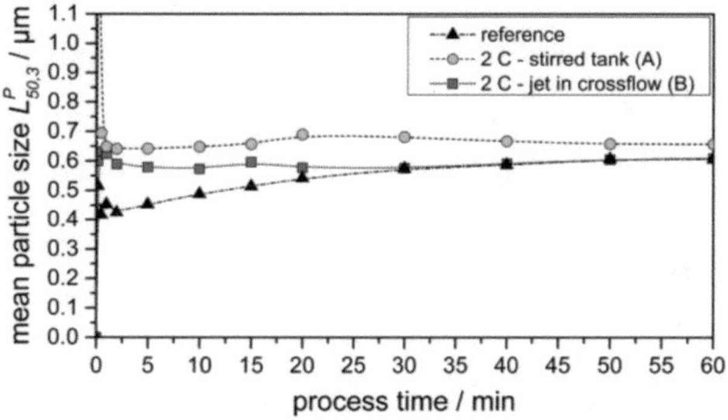

Figure 5.24: Temporal evolution of mean particles size over the process time for the three arrangements investigated (example for $n = 300\ rpm$ and $\dot{V}_{\text{feed}} = 100\ ml/min$).

Regarding the temporal evolutions of mean particle sizes, slight deviations for the first feeding period are observed. At this point, the reasons are not fully clarified. One possibility may be the change in flow conditions for the semi-batch reference vessel with the increasing mixing level, especially at early states of the process (addressed in section 5.5.2.5). Additionally, the data measured in the first $3\text{–}5\ min$ involve a certain uncertainty, since the early particle load of the suspension taken from the $11\ l$ vessel is very low (laser shading of the undiluted sample $< 1\ \%$) and tends to scatter. The longer the feed time, the more representative the PSDs measured become.

Figure 5.25 sums up the results for different stirrer speeds and feed rates comparing the 2C JiC arrangement with the reference. It is evident that the highest congruency can be achieved for experiments at $n = 300\ rpm$, respectively, $\dot{V}_{\text{feed}} = 400\ ml/min$. Less intensive particle sizes predicted for process conditions from the 2C JIC are slightly too large, which implies that the supersaturation predominant in the feed situation is chosen as too slow. Numerical investigations, shown in Figure 5.14, state that a fix $f_{\text{ent}} =$

$\dot{V}_{\text{ent}}/\dot{V}_{\text{prim}} = 0.05$ chosen (necessary design parameter from sections 5.4.2 and 0) is a very simplified assumption.

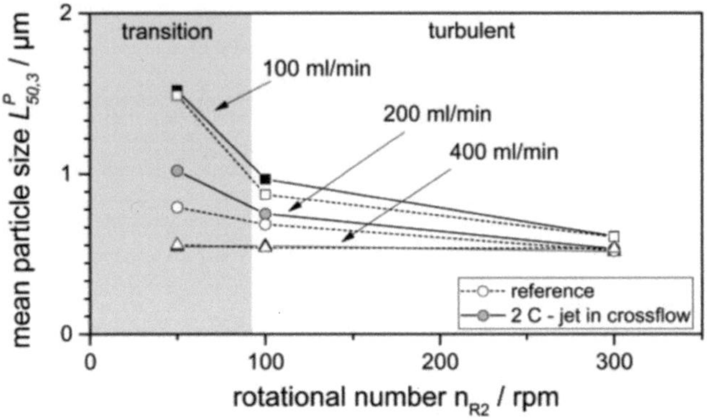

Figure 5.25: Overview of mean particle sizes $L^P_{50,3}$ achieved in the reference case (see Figure 5.22) compared to those from the JiC setup (concept B) for different feed rates and rotational numbers n.

The entrained mass flow that has to be chosen to be circulated by the peristaltic pump depends strongly on the time primary processes need till supersaturation is depleted (τ_{depl}). Figure 5.14 shows that less diluting bulk fluid gets entrained for lower feed rates in a similar period of time. As a result, higher supersaturations are expected, which can also be observed in Figure 5.25. The important of f_{ent} chosen will be further discussed in the flowsheet model section following. However, it is nicely demonstrated that a reference experiment can be reproduced by interconnecting two separated control volumes, which leads to fundamental, simplifying consequences for the flow sheet simulation.

5.7 Results – flowsheet modeling

Constructing on this verified hypothesis, in analogy to section 1, findings for ST reactors are integrated into a flowsheet structure using the framework system *Dyssol* based on C++. Thereby, two approach variations are set up and investigated (see section 5.7.1. and 5.7.2) based on the assumptions made in section 5.3. Important STAR NM findings from the numerical section 5.5.2, such as internal flowrates and entrained mass into the reaction zone, are integrated into the considerations. The general flowsheet procedure using a reaction compartment R_1 and a bulk compartment R_2 can be described as follows (see additionally Figure 5.26).

Certain time points, respectively, numeric time windows, are specified (e.g. $\Delta t = 30s$) within the flowsheet simulation in which the transient semi-batch process is advanced in time. For the sequential modular solution, firstly, the control volume R_1 is balanced, getting the holdup input, for example, composition, from the previous time step. Consonant with the simulation strategy proposed, compartment R_1 also balances the population density of primary processes (identical scheme as Figure 4.1). The population balance has its own internal time stepping interval, as described in section 1, that calculates differentially – at the current specified *Dyssol* timepoint t_{Dyssol} – the actual state of the particle population. After aborting the PB when the supersaturation is depleted, the outlet stream is balanced and handed over to the second control volume R_2, where mass, composition, the number of accumulated particles and the PSD is updated. Thereupon, the next *Dyssol* time step specified by the user can be advanced.

All flowsheet setups are tested for the benchmark case $n = 300\ rpm$ and $\dot{V}_{\mathrm{feed}} = 200\ ml/min$, since, thereby, good experimental conformity between the reference experiment and experimental 2C arrangement (see Figure 5.25) is given.

5.7.1 Variation 1 – two-compartment mixer holdup-assembly

Variation 1 consists of a simplified combination of a CIJM and a holdup tank (Figure 5.26). A feed mass flow $\dot{M}_{\mathrm{feed}}$ of composition $x_{i,\mathrm{feed}}$ is mixed with a certain internal mass flow $\dot{M}_{\mathrm{ent}}$ with the mass fractions x_{i,R_2}, which are entrained into the feed volume and inserted into the reaction zone R_1, as highlighted by the process scheme for R_1 (green box). This internal mass flow $\dot{M}_{\mathrm{ent}}$ is loaded with a fraction of the population density from the holdup R_2 that inherently involves the "history" of the process.

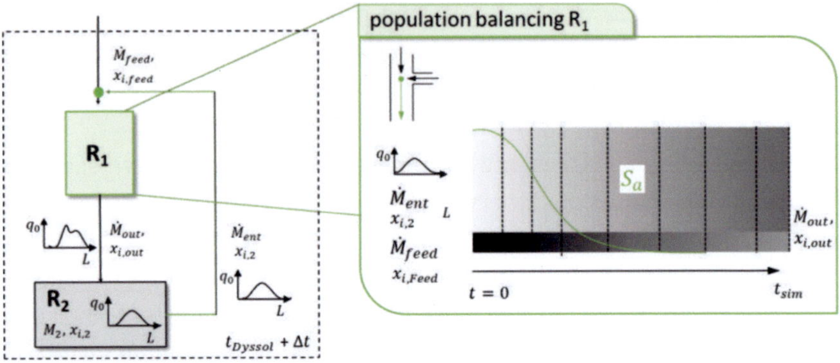

Figure 5.26: Simulation scheme – variation 1, the reaction zone is approximated by a JiC mixer using the CIJM model.

Hereby, the case of instantaneous mixing is discussed (see schematic continuous line for the supersaturation evolution over the simulation time t_{sim} in the green box in Figure 5.26). Similar to the experimental 2C arrangement, the mass flow $\dot{M}_{\text{ent}}$ has to be specified in advance, as arises from the structure of variation 1. The assumption of $f_{\text{ent}} = 0.05$ (eq. 5.15) is made to be in conformity and delivered highly congruent results between 2C JiC and the reference experiments (see 5.6.2).

Figure 5.27, left and right, presents simulation results for the setup presented in Figure 5.26 ($n = 300\,rpm$ and $\dot{V}_{\text{feed}} = 200\,ml/min$). On the one hand, the final simulated PSD q_0 is compared to those gained from experiments, both in the reference vessel and in the 2C JiC setup. Additionally, some selected simulated distributions at different process times are plotted to give an idea of the evolution of particle sizes. The mean particle sizes during the semi-batch process are compared to experiments in Figure 5.27, right.

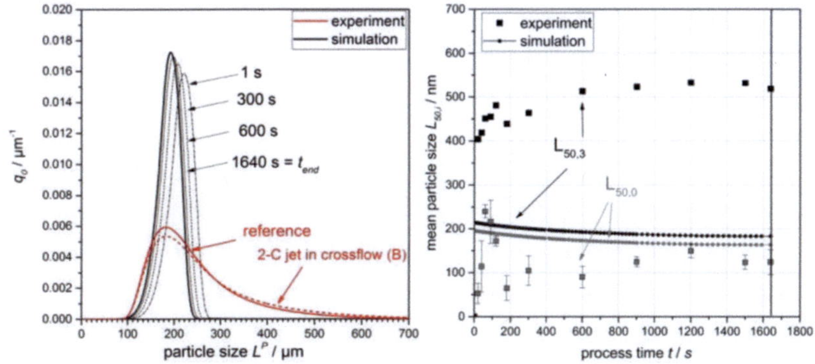

Figure 5.27: Simulation result in comparison to experimental findings. Left: Discrete PSD q_0 compared to simulation results at different time points of the simulation ($n = 300\ rpm$ and $\dot{V}_{\text{feed}} = 200\ ml/min$). Right: Comparison of mean particle sizes.

In order to complete the basis of interpretation, Figure 5.28, left shows the corresponding supersaturation curves at different time steps leading to the PSDs in Figure 5.27, left.

Several interesting conclusions can be drawn. First of all, the simulation predicts the final number of the weighted PSD q_0 fairly well. Thus, mean particle diameters $L^P_{50,0}$ are close to those measured experimentally. However, the dispersion of the particle sizes (see gap of $\sim 400\ nm$ between experimental $L^P_{50,0}$ and $L^P_{50,3}$) cannot be predicted well. This point will be addressed in more detail at the end of chapter 5.7.2.

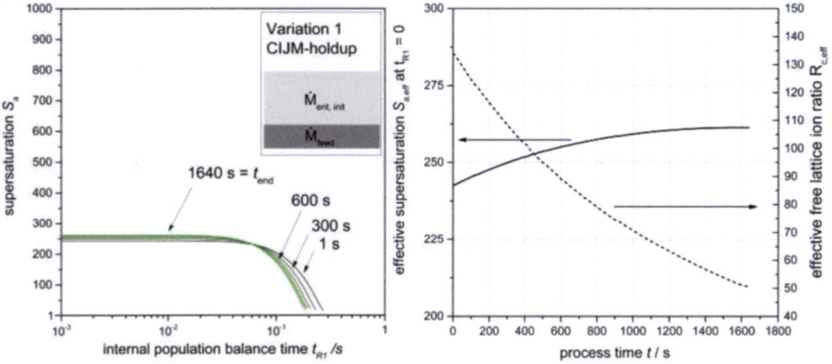

Figure 5.28: Left: Supersaturation progress simulated in R_1 for different time points in the semi-batch process ($n = 300\ rpm$ and $\dot{V}_{\text{feed}} = 200\ ml/min$).

One further interesting point is that the supersaturation during the process increases, although the overall concentration of Ba^{2+} ions in the tank is decreasing over time (see also Figure 5.31). This leads to decreasing particle sizes along the process time. Figure 5.28, right, gives an unexpected but simple explanation of that behavior. Initially, precipitation is carried out strongly nonstoichiometrically ($\mathrm{R_{c,init}} = 135$). In the process, $\mathrm{R_c}$ decreases, since barium ions are consumed. This leads to an increasing supersaturation S_a due to the nonlinear behavior of the activity coefficient model (Pitzer, see Figure 2.4). Surprisingly, such a decreasing behavior in particle size in the experimental section (Figure 5.24) is also found for the 2C JiC approach that is generically very similar to this flowsheet simulation. At this point, this analogy between the two findings is not fully confirmed, but may come from the same source. This size-decreasing effect is not observed in the reference experiment (Figure 5.27). Thus, slight mechanistic differences between the reference and 2C arrangement might exist which will be discussed in the next section in "variation 2."

A third point that should be pointed out is that supersaturation is depleted in our simulations after 0.2 to $0.3\ s$, which supports the design assumption made for the experimental setup nicely. On the one hand, the reference tank is large enough to decouple primary and secondary processes, on the other hand, the residence times in the reaction compartments $\mathrm{R_1}$ are dimensioned long enough.

5.7.2 Variation 2 – two-compartment jet in crossflow

Variation 2 is an approach enhancement which overcomes the crucial problem of the need to specify the recirculation stream $\dot{M}_{\mathrm{ent}}$ in advance (as done in the "experimental simulation" and in "variation 1"). This quantity represents the problematic parameter for a correct description of the mixing situation ($\dot{M}_{\mathrm{ent}} = f(\tau_{\mathrm{depl}})$). Since τ_{depl} is initially unknown, the question regarding how much bulk fluid is involved in mixing is similarly open (discussed in section 5.5.2). Thus, a new approach is presented which allows for a flexible reaction compartment size $\mathrm{R_1}$. Such a procedure is much more generally valid and increases the scalability of the flow sheet tool significantly. Figure 5.29 shows the interconnection of the two control volumes R_1 and R_2 schematically. In contrast to variation 1, the mass flow added, $\dot{M}_{\text{ent}} = f(\tau)$ as the missing information, is taken from CFD (STAR NM) and added in an empirical form. Such CFD "measurements" are presented in Figure 5.14.

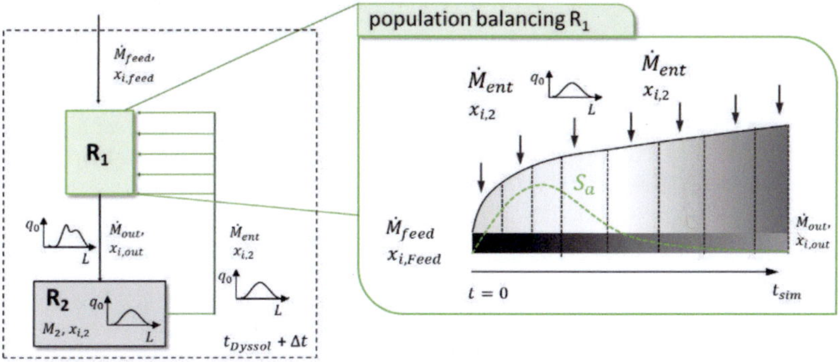

Figure 5.29: Simulation scheme, variation 2, flexible arrangement, generically close to the processes occurring in the ST reactor.

It is assumed for the flowsheet simulation executed that the first slope in the entrained mass flow rates measured (expressed empirically by eq. 5.16) dominates the process. However, it should be considered that the rate of entrainment clearly decreases when the fluid reaches the vessel wall and, thus, is exposed to lower energy dissipation rates. However, eq. 5.16 should be a good approximation for the simulation presented, since depletion times are supposed to stay in the quasi-linear section observed (see Figure 5.32). Two options are discussed in the following. An auxiliary scheme is given in Figure 5.30 to recognize the differences of option 1 and option 2.

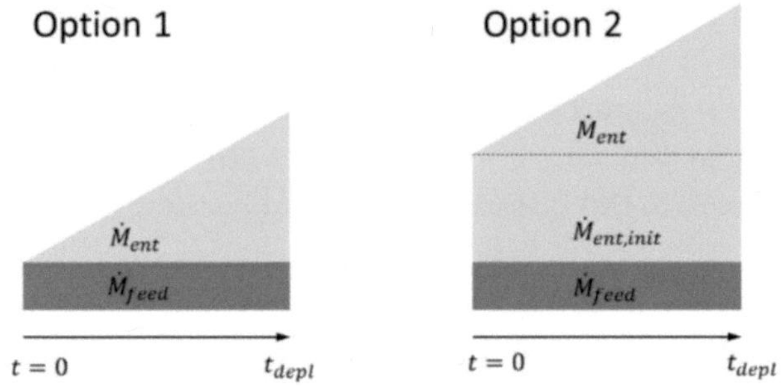

Figure 5.30: Scheme of the options investigated. On the one hand, entrainment solely following eq. 5.16 is considered. On the other hand, a premixed fraction is assumed to adjust $\langle \bar{S}_{a,init} \rangle$ (Figure 5.19) at the mixing point.

Option 1: $\dot{V}_{\mathrm{ent}}(t = 0) = 0$

Mixing processes in R_1 are simulated as physically assumed in option 1. The unmixed feed is brought initially into contact with the entrained bulk volume. Thereby, entrainment follows the rate (simplified first-order approximation, see eq. 5.16) extracted from the CFD (see 5.5.2.1). An infinitesimal value $\dot{V}_{\mathrm{ent}}(t = 0) = 10^{-6} \cdot \dot{V}_{\mathrm{feed}} \approx 0$ is preset to provide for initial solver problems.

Option 2: $\dot{V}_{\mathrm{ent}}(t = 0) = \dot{V}_{\mathrm{ent,init}}(STAR\ NM)$

In option 2, the initial supersaturation value "measured" within the CFD (extrapolated to t = 0, ($\langle \bar{S}_{\mathrm{a,init}} \rangle$, see Figure 5.19, dashed lines) is preset by adding a premixed fraction $\dot{V}_{\mathrm{ent,init}}$. This mixing point ($S_a$ = 330) can be reached by choosing $\dot{V}_{\mathrm{ent}}(t = 0) = 18 \cdot \dot{V}_{\mathrm{feed}}$ for the example demonstrated ($n = 300$ and $\dot{V}_{\mathrm{feed}}$ = 200 ml/min). For this case, $f_{\mathrm{ent}}(t = 0) = 0.048$ (see "option 2" in Figure 5.31, right at t = 0), which is very interesting, since this value fits the parameter chosen experimentally and which delivered high congruency between the reference experiment and the 2C JiC arrangement. Starting from the initial composition, the fluid explained, similar to option 1, is entrained according to the function in eq. 5.16.

Figure 5.31, left, shows the molar concentration $\tilde{c}_i$ of the components involved in the holdup vessel R₂, simulated either for option 1 or option 2. (By the way, this diagram is also valid for "variation 1".) Primary processes are terminated in the reaction compartment R_1, as required within the simulation strategy (section 5.3). This means that the sulfate concentration fed stays in R_2 below the solubility product $K_{sp,\mathrm{BaSO_4}}(\approx 0)$, since it is fully consumed by particle formation in the reaction zone R_1.

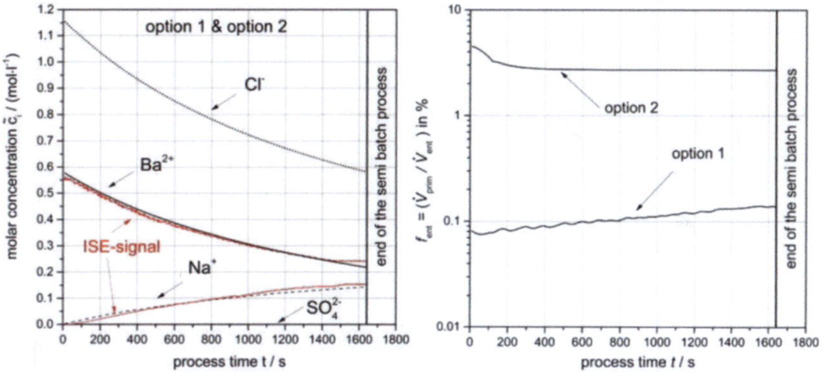

Figure 5.31: Left: Concentration evolution in the holdup R_2, right: $f_{ent}(t)$ for $n = 300\ rpm$ and $\dot{V}_{\mathrm{feed}} = 200\ ml/min$, option 1: Initially unmixed feed, and option 2: Premixed feed with entrainment versus the semi-batch process time.

The barium ions reduce due to dilution and consumption during the particle formation. The experimentally measured, calibrated signals of two-ion-selective electrodes, ISE ($\rightarrow Ba^{2+}$ – signal = reactive component and Na^{+} – signal = inert component) measured within the ST reactor are added in Figure 5.31, left, to confirm this model behavior. The electrodes are placed at the opposite radial position of the feed pipe. Thereby, it can be concluded that the assumption of segregated zones R_1 and R_2 is permitted and that the global balancing of the flow sheet model works correctly. Findings presented are made for all experiments investigated (5.6.1).

Moreover, Figure 5.31, right, shows the factor $f_{\mathrm{ent}} = f(t)$ for the two simulation options. When interpreting these curves, Figure 5.32 is very helpful, since it shows the supersaturation depletion curves in the reaction compartment R_1 at different time points of process times. On the one hand, in option 1, only a small amount of fluid is entrained until the supersaturation is depleted, since particle formation is fast due to the relatively high maximal values of supersaturation reached. Throughout the process time, the quantity of entrained fluid increases, since the maximal supersaturation reached decreases and, thus, particle formation time decreases (see Figure 5.32, left). On the other hand, in option 2, the amount of entrained fluid is due to the model specification being higher. Supersaturation takes place (as observed within STAR NM) at lower values than those reached within option 1. Moreover, the fractional amount of entrained fluid f_{ent} decreases, since the depletion velocity increases significantly due to the entrained particles which offer an additional surface for growth (see Figure 5.32, right).

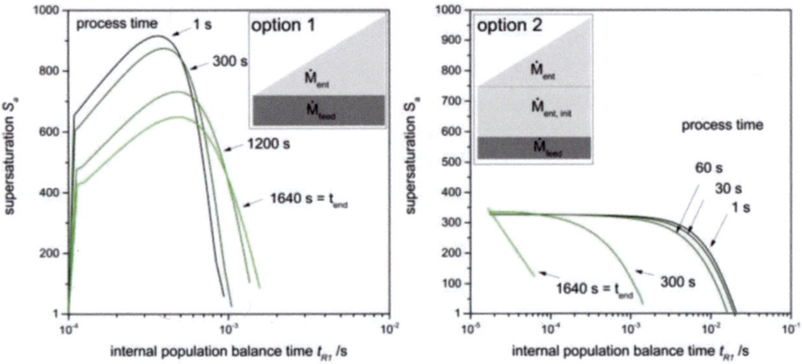

Figure 5.32: Supersaturation profiles calculated for R_1 at different process times; left: Option 1, right: Option 2 ($n = 300\,rpm$ and $200\,ml/min$).

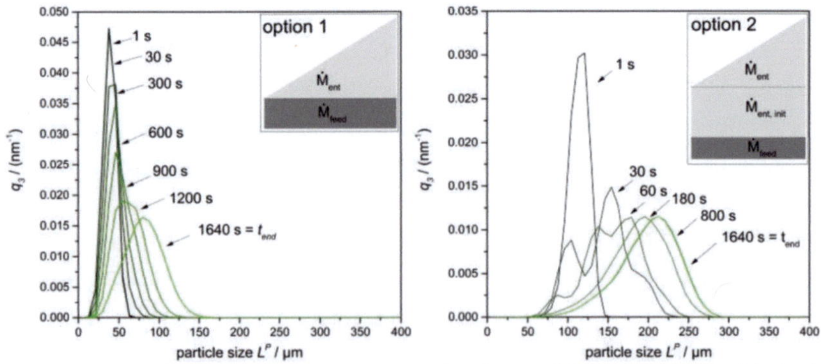

Figure 5.33: Particle size evolution over process time for $n = 300\,rpm$ and $200\,ml/min$, left: Option 1, initially unmixed feed; right: Partially premixed feed according to the extrapolated value from STAR NM (see Figure 5.19).

These significantly different courses of supersaturation also have a dramatic impact on the particle sizes calculated within the population balance (see Figure 5.33). Figure 5.33 shows PSDs q_3 calculated throughout the process times for option 1 (left) and option 2 (right). Option 1 tends to generate significantly smaller particles than option 2. Moreover, fine disperse particles are formed in the first few minutes of process time. Bigger particles arise later in the process. Option 2, by contrast, has the important changes in PSD in the first 1 to 3 min, and then, no longer changes significantly (see also the progress of simulated mean particle sizes in Figure 5.34).

Important statements can be made regarding the experiments. Option 1 (only considering entrainment) does not reflect the experimental behavior well and does not seem to be an adequate process description. Simulations with the flowsheet model option 2, considering the full CFD received STAR NM information with a certain amount of premixed fluid, is in good accordance with the experiments. Not only is the final mean particle size predicted fairly well, moreover, the trend that particle size changes mainly over the first few minutes of process time can be confirmed by experiments.

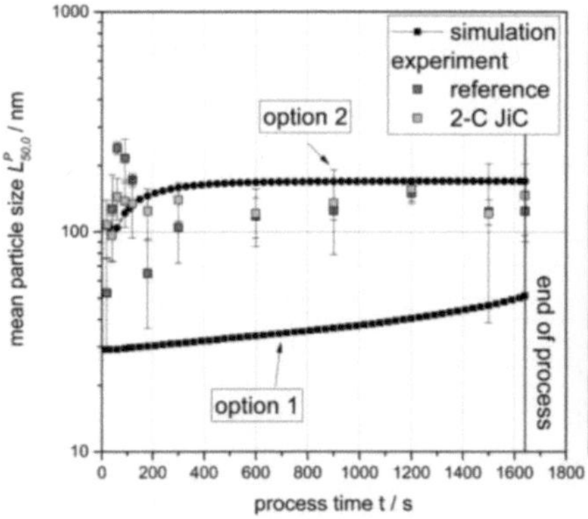

Figure 5.34: Left: Mean particle size evolution $L^P_{50,0}$ over time, option 1 and option 2 compared to experimental data measured with the Zetasizer. Right: SEM images of particles for $n = 300\ rpm\ \dot{V}_{feed} = 200\ ml/min$.

However, it should be mentioned, as also observed for variation 1 (section 5.6.2), that the $L^P_{50,0}$ is predicted well, but the mean volumetric particle sizes

$L^P_{50,3}$ presented, for example, in Figure 5.22 are clearly deviating. Figure 5.35 illustrates this problem. In a similar way to the predictions of the CIJM model in the laminar regime (Figure 3.37), the simulated PSD is significantly narrower than the ones from experiments.

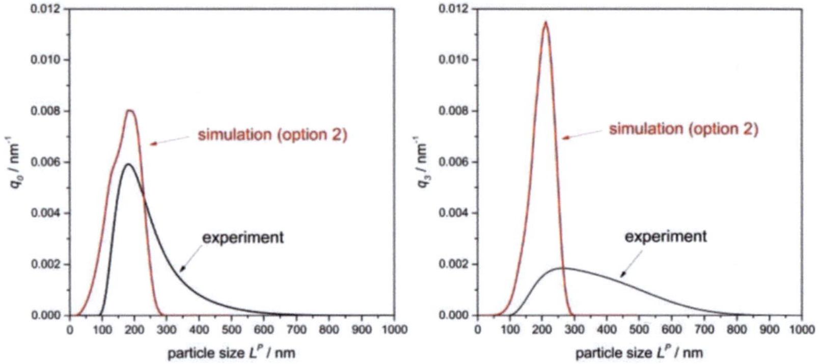

Figure 5.35: Discrete PSD of the experimentally measured and the simulated PSD (option 2); left: q_0, right: q_3.

Dispersion, which is created by the pulsating discharge flow of the impeller (→temporal dependencies, see Figure 5.18) or which exists in any way in the "measurement" plane (→spatial dependencies, Figure 5.18), is not considered at this stage of the approach development. Since the nucleation rate is strongly nonlinear (Figure 2.8), lower supersaturated regions can lead to significantly bigger particles.

However, this section demonstrates how numerical coarse-graining complemented with a multicompartment structure enable the creation of an understanding of complex process interactions. Further thinking on how to take into account the dispersion of particle sizes should be made in the future.

6 Conclusion and Outlook

The superordinate aim of this work is to setup a predictive scale-comprehensive flowsheet model for the process simulation of fast precipitating systems. Thereby, the thesis focuses basically on the broad material class of inorganic solids which are formed out of the aqueous phase. Precipitation crystallization, especially for the nanoparticle synthesis, is one of the most important unit operations. The particle sizes required which lead, for example, to a preferable bioavailability in medicine, to good coating properties for paints or to desired attributes for a catalyst, are adjusted mostly by trial and error. Process simulation can be one key element to help one understand the prevailing multiscale phenomena, respectively, time and space, and the interplay of frequently superimposed mechanisms of solid formation, and to design a process precisely and its final properties.

The main features such a process simulation should take into account are the complex hydrochemistry of concentrated ionic mixtures, fundamental kinetics of solid formation that are in a time range that brings experimental access to its limits (~ms) and the influence of mixing that can, depending on the apparatus used, influence the particle formation strongly. The $BaSO_4$ precipitation from solvated Na_2SO_4 and $BaCl_2$ reactants is taken as a well characterized model system. Two apparatuses are investigated. On the one hand, method development is carried out using CIJMs, which mainly find their application in laboratory use. On this basis, a first method transfer on ST reactors, which are the most common precipitation devices in industrial application, is presented. Therefore, new scale-combining modeling concepts and balancing ideas, such as a multicompartment approach, are presented.

Throughout this work, CFD is considered to be the important tool that helps one to gain insight into temporally inaccessible procedures and to "measure" quantities, such as local supersaturations, that are not measurable experimentally.

6.1 Method development – confined impinging jet mixer

In this thesis, the CIJM is chosen as the important benchmark apparatus for the method development. The influence of mixing in these devices can be decreased gradually by increasing the flow rate until a plateau in the process performance is reached where mixing no longer plays a role. Thus, an experimental database and a comparison to numerical results can be established highly representatively. Moreover, the fundamental mechanism of precipitation can be studied initially without mixing influence. This means the complexity of the model can be increased step-by-step. In the following, important investigations and advances for CIJMs are made leading finally to a predictive flow sheet tool.

A CFD tool is established initially that enables numerical "measurements" close to reality. Therefore, an experimentally reliable database on CIJMs is created by investigating, for example, the pressure drop in the mixing chamber, the depletion of segregation by LIF and the velocity field by PIV. In accordance with literature, dominant flow regimes are identified which can be confirmed numerically by the 3D-detached eddy SST-k-ω simulation chosen, using the species transport model. Moreover, it can be shown that the mean particle sizes at precipitation and the dispersion of the product distribution are strongly dependent on the adjusted flow regime.

In this work, one focus is placed on the investigation of mixing phenomena on the particle formation process. This influence is strongly dependent on the apparatus used (apparatus function). The prediction of mixing influences is not only valid for precipitation for the models worked out, but also for other swift processes, such as fast parallel reactions. Approaches from different degrees of detail are investigated to tackle this mixing topic. Thus, a fully resolved CFD method including a moment-based population balance approach (DQMOM), a CFD coarse-graining method and, more fundamental, mixing theory-based approaches are tested regarding their suitability. One of the main advancements this thesis provides is the new coarse-graining approach mentioned, which is called STAR NM (“spatially and temporally averaged reduced numeric measurement”). This method consists of the interplay of “fast” single phase 3D CFD calculations, “measurements” of specific process-dominating state variables (e.g. supersaturation) and their use by a 1D population balance. The method allows the gainful reduction of the computational load compared to the fully resolved approach. Extensive CFD information is diminished to a reasonable level, whereas the particle sizes predicted are in high accordance with experiments. The method allows one to show that the characteristic shape of the PSD (tailing) originates from the residence time spectrum in the mixer.

A further advancement this thesis provides is the proposal of a scale-up hypothesis for CIJMs. Mixing times from STAR NM are, on the one hand, used to confirm temporal predictions of mixing approaches based on the classic mixing theory. Moreover, it can be demonstrated how mesomixing phenomena (dispersion of the fresh feed, eddy disintegration) slow down the mixing speed for increasing mixing geometries. This matter is of fundamental importance, since it gives a precise explanation of why the scale-up of CIJMs has not yet been understood.

In a final step, a flow sheet module for CIJMs is set up which can handle influences of the material function (e.g. different materials or different supersaturations adjusted, which are, in a first instance, independent of the apparatus) and the influence of the apparatus function (mixing influence, scale-up) of the CIJM. The module possesses a wide degree of freedom: for example, the crucial question of how to implement the complex hydrochemistry in a more general form is solved by integrating the hydrochemistry software *PhreeqC*, which offers broad databases. Further benchmark tests, for example, for additional material systems or CIJMs of industrial scale, would be beneficial to evaluate the predictive power of the tool. One important aspect that should be considered carefully is the particle synthesis sometimes influenced by aggregative or dendritic growth, which currently outlines the limits of the approach presented.

6.2 Method transfer – stirred-tank reactor

In a further step, a method transfer regarding ST reactors is advanced. One central question considering the methodological toolbox created for CIJMs is their transferability to a system which is inherently more complex due to, for example, recirculation streams or strongly inhomogeneous flow fields.

Global balancing ideas from literature are collected and advanced with my own approaches to overcome the fundamental problem that precipitation in an ST reactor is hard to describe. Dominating scales at precipitation (temporally and spatially) are significantly smaller than the macroscale of an industrial vessel. Therefore, an interconnected compartment approach is proposed which can resolve micro- and mesoscales affecting the solid formation reasonably. Thus, a 2C approach with a reaction zone R_1 and a bulk zone R_2 is considered for the semi-batch process with a feeding position close to the stirrer.

In a first important step in this work, it is demonstrated experimentally that multicompartment approaches, in this case a 2C arrangement, are a reasonable abstraction for fast precipitating systems in ST reactors. Thereby, an 11-l ST with a Rushton turbine is used for the investigation. It is found that the JiC situation (concept B, section 5.6.2) is supposed to be a generally well-suited approximation for the feed situation in the discharge of the impeller. Thereby, the most critical parameter that must be adjusted during experiments is the recirculation flow $\dot{V}_{\mathrm{ent}}$ (specified by the factor $f_{\mathrm{ent}} = \dot{V}_{\mathrm{ent}}/\dot{V}_{\mathrm{prim}}$), which affects the composition directly, and the supersaturation at the mixing point. In this work, in a first assumption supported by CFD considerations, a fixed value of $f_{\mathrm{ent}} = 0.05$ has been chosen and found to be a first reasonably good choice for most of the experiments carried out. In reality, this factor is dependent on the feed pipe diameter, the feed rate and the stirrer speed, and certainly varies over the process time, which should be considered. Flow sheet simulations subsequently verify this issue.

An important advancement that is realized for the interconnected compartments in the flowsheet module is the adaptive size of the reaction compartment R_1 for the time points simulated. Bulk fluid is entrained as long as supersaturation exists. This feature is an essential improvement, since the mixed bulk mass flow does not have to be specified in advance. For the semi-batch process regarded, this improvement leads to a promising scale-up potential for technical precipitation reactors in the scale of $\sim m^3$. If the entrainment rate and the initial supersaturation at the feed inlet are known (from CFD), the approach can handle any vessel size.

At this point, how to transfer the STAR NM methodology developed for the CIJMs on ST reactors is demonstrated successfully. Thereby, supersaturations measured by CFD and entrained mass flow rates are spatially and temporally averaged and handed over to the decoupled 1D population balance in a flow sheet model. The flow sheet model (option 2) discussed, involving a certain premixed feed and an entrainment rate throughout the particle formation time, delivered the first congruent results of selected experiments. Thereby, it can be determined that the initial mixing state and the initial supersaturation ($S_{a,\text{init}}$, see Figure 3.19) dominates particle formation strongly. Moreover, most dynamic changes in the PSD are found to occur at the beginning of the semi-batch process. It is fair to say that the full dispersion of the PSD cannot be predicted currently with this flow sheet model. Similar problems are also found in the CIJMs regarding laminar mixing. A sectional averaging may be able to increase the accuracy of STAR NM.

It can be stated that, as an outlook, a mixing theory-based description of the entrained flowrates (eq. 6.1), similar to the LTSA for the CIJM, would represent an advancement to create a universal ST module. Mesomixing from a scale Λ_c and micromixing effects mainly depending on ε must be taken into account. Effects of the feed pipe diameter, for instance, could be considered in such a manner.

$$\frac{\partial \dot{M}_{\text{ent}}}{\partial t} = f(\Lambda_c, \varepsilon) \qquad 6.1$$

The course of the energy dissipation ε along a streamline of freshly fed feed can be considered, for example, including the findings from the numerical section (5.5.2.3).

Regarding the feed situation close to the impeller, eq. 5.19 combined with 5.20, investigated within this work, should be a good approximation to describe the course of energy dissipation a streamline of fresh feed undergoes. Instead of a mean value, as used in the CIJM, a temporal progress should be considered. A description of mesoscale Λ_c (eq. 6.2) evolving in time in dependency of the energy dissipation is suggested by Baldyga et al. [128].

$$\frac{\partial \Lambda_c}{\partial t} = C\varepsilon^{1/3}{\Lambda_c}^{1/3}$$

$$\text{with } \Lambda_c(t=0) = \sqrt{\dot{V}_{\text{feed}}/(\pi \cdot u_{\text{loc}})} \qquad 6.2$$

7 References

[1] O. Söhnel and J. Garside, *Precipitation: Basic principles and industrial applications.* Oxford: Butterworth-Heinemann, 1992.

[2] J. Gradl, H.-C. Schwarzer, F. Schwertfirm, M. Manhart, and W. Peukert, "Precipitation of nanoparticles in a T-mixer: coupling the particle population dynamics with hydrodynamics through direct numerical simulation," *Chem. Eng. Process.: Process Intens.*, vol. 45, no. 10, pp. 908–916, 2006.

[3] R. T. Kügler, S. Doyle, and M. Kind, "Fundamental insights into barium sulfate precipitation by time-resolved in situ synchrotron radiation wide-angle X-ray scattering (WAXS)," *Chem. Eng. Sci.*, vol. 133, pp. 140–147, 2015.

[4] M. Kucher and M. Kind, "Vom Keim zum Partikel: Experimentelle und numerische Untersuchungen zur Partikelbildung bei der Fällung von Bariumsulfat," *Chem. Ing. Tech.*, vol. 79, no. 9, pp. 1396–1397, 2007.

[5] H.-C. Schwarzer, "Nanoparticle precipitation: an experimental and numerical investigation including mixing," Ph.D. dissertation, FAU Erlangen-Nürnberg, 2005.

[6] C. Steyer, "Precipitation of barium sulfate in a semi-batch stirred tank reactor: Influence of feeding policy on particle size and morphology," Ph.D. dissertation, Magdeburg, Universität, 2012.

[7] L. Vicum and M. Mazzotti, "Multi-scale modeling of a mixing-precipitation process in a semibatch stirred tank," *Chem. Eng. Sci.*, vol. 62, no. 13, pp. 3513–3527, 2007.

[8] A. Mersmann, *Crystallization technology handbook*. CRC Press, 2001.

[9] R. J. LeVeque, *Finite volume methods for hyperbolic problems*. Cambridge University Press, 2002.

[10] D. Ramkrishna and M. R. Singh, "Population balance modeling: current status and future prospects," *Annu. Rev. Chem. Biomol. Eng.*, vol. 5, pp. 123–146, 2014.

[11] Z. Li, *Fortgeschrittene Methoden zur Beschreibung der Wirbelschicht-Sprühgranulation*. KIT Scientific Publishing, 2013.

[12] R. Gunawan, I. Fusman, and R. D. Braatz, "High resolution algorithms for multidimensional population balance equations," *AIChE J.*, vol. 50, no. 11, pp. 2738–2749, 2004.

[13] K. W. Morton and D. F. Mayers, *Numerical solution of partial differential equations*. Cambridge University Press, 1994.

[14] D. L. Marchisio, J. T. Pikturna, R. O. Fox, R. D. Vigil, and A. A. Barresi, "Quadrature method of moments for population-balance equations," *AIChE J.*, vol. 49, no. 5, pp. 1266–1276, 2003.

[15] R. McGraw, "Description of aerosol dynamics by the quadrature method of moments," *Aerosol Sci. Technol.*, vol. 27, no. 2, pp. 255–265, 1997.

[16] D. L. Marchisio and R. O. Fox, "Solution of population balance equations using the direct quadrature method of moments," *J. Aerosol Sci*, vol. 36, no. 1, pp. 43–73, 2005.

[17] R. Fan, N. Xie, F. Battaglia, and R. O. Fox, "A reference guide to use MFIX with ISAT and DQMOM," *National Energy Technology Laboratory, Archival Reports.*, 2004.

[18] Y. Liu and R. O. Fox, "CFD predictions for chemical processing in a confined impinging-jets reactor," *AIChE J.*, vol. 52, no. 2, pp. 731–744, 2006.

[19] K. Schaber, *Scriptum: Chemische Thermdynamik*, I. for Technical Thermodynamics and R. (ITTK), Karlsruhe Institute of Technology, 2010.

[20] J. W. Mullin, *Crystallization*. Butterworth-Heinemann, 2001.

[21] M. Kucher, D. Babic, and M. Kind, "Precipitation of barium sulfate: experimental investigation about the influence of supersaturation and free lattice ion ratio on particle formation," *Chem. Eng. Process.: Process Intens.*, vol. 45, no. 10, pp. 900–907, 2006.

[22] L. A. Bromley, "Thermodynamic properties of strong electrolytes in aqueous solutions," *AIChE J.*, vol. 19, no. 2, pp. 313–320, 1973.

[23] K. S. Pitzer, "Thermodynamics of electrolytes. I. Theoretical basis and general equations," *J. of Phys. Chem.*, vol. 77, no. 2, pp. 268–277, 1973.

[24] C. Steyer, M. Mangold, and K. Sundmacher, "Modeling of particle size distribution for semibatch precipitation of barium sulfate using different activity coefficient models," *Ind. Eng. Chem. Res.*, vol. 49, no. 5, pp. 2456–2468, 2010.

[25] C. W. Davies, *Ion association*. Butterworths, 1962.

[26] A. H. Truesdell and B. F. Jones, "WATEQ, a computer program for calculating chemical equilibria of natural waters," *J. Res. US Geol. Surv*, vol. 2, no. 2, pp. 233–248, 1974.

[27] A. Felmy, D. Rai, and J. Amonette, "The solubility of barite and celestite in sodium sulfate: evaluation of thermodynamic data," *J. Solution Chem.*, vol. 19, no. 2, pp. 175–185, 1990.

[28] C. Monnin, "A thermodynamic model for the solubility of barite and celestite in electrolyte solutions and seawater to 200 C and to 1 kbar," *Chem. Geol.*, vol. 153, no. 1, pp. 187–209, 1999.

[29] L. Vicum, M. Mazzotti, and J. Baldyga, "Applying a thermodynamic model to the non-stoichiometric precipitation of barium sulfate," *Chem. Eng. Technol.*, vol. 26, no. 3, pp. 325–333, 2003.

[30] L. Metzger and M. Kind, "The influence of mixing on fast precipitation processes–A coupled 3D CFD-PBE approach using the direct quadrature method of moments (DQMOM)," *Chem. Eng. Sci.*, 2016.

[31] D. Kashchiev, *Nucleation*. Butterworth-Heinemann, 2000.

[32] K. Schaber, *Scriptum: Thermodynamik disperser Systeme*, I. for Technical Thermodynamics and R. (ITTK), Karlsruhe Institute of Technology, 2015.

[33] A. Mersmann and M. Kind, "Chemical engineering aspects of precipitation from solution," *Chem. Eng. Technol.*, vol. 11, no. 1, pp. 264–276, 1988.

[34] A. E. Nielsen and O. Söhnel, "Interfacial tensions electrolyte crystal-aqueous solution, from nucleation data," *J. Cryst. Growth*, vol. 11, no. 3, pp. 233–242, 1971.

[35] O. Söhnel, "Electrolyte crystal-aqueous solution interfacial tensions from crystallization data," *J. Cryst. Growth*, vol. 57, no. 1, pp. 101–108, 1982.

[36] O. Söhnel, "Estimation of electrolyte-crystal-aqueous-solution interfacial tension," *J. Cryst. Growth*, vol. 63, no. 1, pp. 174–176, 1983.

[37] S. Li, J. Xu, and G. Luo, "Control of crystal morphology through supersaturation ratio and mixing conditions," *J. Cryst. Growth*, vol. 304, no. 1, pp. 219–224, 2007.

[38] D. Gebauer and H. Cölfen, "Prenucleation clusters and non-classical nucleation," *Nano Today*, vol. 6, no. 6, pp. 564–584, 2011.

[39] R. Taylor and R. Krishna, *Multicomponent mass transfer*. John Wiley & Sons, 1993.

[40] E. Samson and J. Marchand, "Numerical solution of the extended Nernst–Planck model," *J. Colloid Interface Sci.*, vol. 215, no. 1, pp. 1–8, 1999.

[41] D. A. Walker, "A fluorescence technique for measurement of concentration in mixing liquids," *J. Phys. E: Scientific Instruments*, vol. 20, no. 2, p. 217, 1987.

[42] H.-C. Schwarzer and W. Peukert, "Experimental investigation into the influence of mixing on nanoparticle precipitation," *Chemical engineering & technology*, vol. 25, no. 6, pp. 657–661, 2002.

[43] J. Schlomach, *Feststoffbildung bei technischen Fällprozessen: Untersuchungen zur industriellen Fällung von Siliziumdioxid und Calciumcarbonat*. Univ.-Verlag Karlsruhe, 2006.

[44] H. Oertel jr, M. Böhle, and T. Reviol, *Übungsbuch Strömungsmechanik: Grundlagen, Grundgleichungen, Analytische und Numerische Lösungsmethoden*. Springer-Verlag, 2012.

[45] J. Fröhlich and D. von Terzi, "Hybrid LES/RANS methods for the simulation of turbulent flows," *Prog. Aerosp. Sci.*, vol. 44, no. 5, pp. 349–377, 2008.

[46] V. Ramesh, S. Vengadesan, and J. L. Narasimhan, "3D unsteady RANS simulation of turbulent flow over bluff body by non-linear model," *Int. J. Numer. Methods Heat Fluid Flow*, vol. 16, no. 6, pp. 660–673, 2006.

[47] M. Strelets, "Detached eddy simulation of massively separated flows," in *AIAA, Aerospace Sciences Meeting and Exhibit, 39 th, Reno, NV*, 2001.

[48] F. R. Menter, M. Kuntz, and R. Langtry, "Ten years of industrial experience with the SST turbulence model," *Turbulence, Heat Mass Tansfer.*, vol. 4, no. 1, pp. 625–632, 2003.

[49] M. S. Gritskevich, A. V. Garbaruk, J. Schütze, and F. R. Menter, "Development of DDES and IDDES formulations for the k-ω shear stress transport model," *Flow, Turbul. Combust.*, vol. 88, no. 3, pp. 431–449, 2012.

[50] P. R. Spalart, S. Deck, M. L. Shur, K. D. Squires, M. K. Strelets, and A. Travin, "A new version of detached-eddy simulation, resistant to ambiguous grid densities," *Theor. and comp. fluid dyn.*, vol. 20, no. 3, pp. 181–195, 2006.

[51] M. Manninen, V. Taivassalo, S. Kallio *et al.*, "On the mixture model for multiphase flow," *VTT Pub. 288*, 1996.

[52] B. K. Johnson and R. K. Prud'homme, "Chemical processing and micromixing in confined impinging jets," *AIChE J.*, vol. 49, no. 9, pp. 2264–2282, 2003.

[53] A. J. Mahajan and D. J. Kirwan, "Micromixing effects in a two-impinging-jets precipitator," *AIChE J.*, vol. 42, no. 7, pp. 1801–1814, 1996.

[54] L. Metzger and M. Kind, "On the transient flow characteristics in Confined Impinging Jet Mixers-CFD simulation and experimental validation," *Chem. Eng. Sci.*, vol. 133, pp. 91–105, 2015.

[55] S. W. Siddiqui, Y. Zhao, A. Kukukova, and S. M. Kresta, "Characteristics of a confined impinging jet reactor: energy dissipation, homogeneous and heterogeneous reaction products, and effect of unequal flow," *Ind. Eng. Chem. Res.*, vol. 48, no. 17, pp. 7945–7958, 2009.

[56] H. D. Laufhütte and A. Mersmann, "Local energy dissipation in agitated turbulent fluids and its significance for the design of stirring equipment," *Chem. Eng. Technol.*, vol. 10, no. 1, pp. 56–63, 1987.

[57] V.-G. V. und Chemieingenieurwesen (Ed.). VDI Heat Atlas, *chap. L2.2*. Springer, Heidelberg/Dordrecht/London/New York, 2010.

[58] R. J. Santos and M. A. Sultan, "State of the art of mini/micro jet reactors," *Chem. Eng. Technol.*, vol. 36, no. 6, pp. 937–949, 2013.

[59] M. A. Sultan, D. M. M. Fonte, C. P., J. C. B. Lopes, and R. J. Santos, "Experimental study of flow regime and mixing in T-jets mixers," *Chem. Eng. Sci.*, vol. 73, pp. 388–399, 2012.

[60] N. Dombrowski and W. R. Johns, "The aerodynamic instability and disintegration of viscous liquid sheets," *Chem. Eng. Sci.*, vol. 18, no. 3, pp. 203–214, 1963.

[61] P. Danckwerts, "The definition and measurement of some characteristics of mixtures," *Appl. Sci. Res. A*, vol. 3, no. 4, pp. 279–296, 1952.

[62] A. M. Teixeira, R. J. Santos, M. R. P. Costa, and J. C. B. Lopes, "Hydrodynamics of the mixing head in RIM: LDA flow-field characterization," *AIChE J.*, vol. 51, no. 6, pp. 1608–1619, 2005.

[63] D. Bothe, A. Lojewski, and H.-J. Warnecke, "Computational analysis of reactive mixing in T-microreactors," in *Micro and Macro Mixing, H. Bockhorn, D. Mewes, W. Peukert, and H-J. Warnecke (eds.)*. Springer, 2010, pp. 265–286.

[64] S. C. Malguarnera and N. P. Suh, "Liquid injection molding I. An investigation of impingement mixing," *Polym. Eng. Sci.*, vol. 17, no. 2, pp. 111–115, 1977.

[65] C. L. Tucker and N. P. Suh, "Mixing for reaction injection molding. I. Impingement mixing of liquids," *Polym. Eng. Sci.*, vol. 20, no. 13, pp. 875–886, 1980.

[66] E. Gavi, L. Rivautella, D. L. Marchisio, M. Vanni, A. Barresi, and G. Baldi, "CFD modelling of nano-particle precipitation in confined impinging jet reactors," *Chem. Eng. Res. Des.*, vol. 85, no. 5, pp. 735–744, 2007.

[67] G. di Veroli and S. Rigopoulos, "Modeling of turbulent precipitation: A transported population balance-PDF method," *AIChE J.*, vol. 56, no. 4, pp. 878–892, 2010.

[68] C. Lindenberg and M. Mazzotti, "Experimental characterization and multi-scale modeling of mixing in static mixers. Part 2. Effect of viscosity and scale-up," *Chem. Eng. Sci.*, vol. 64, no. 20, pp. 4286–4294, 2009.

[69] D. R. Unger and F. J. Muzzio, "Laser-induced fluorescence technique for the quantification of mixing in impinging jets," *AIChE J.*, vol. 45, no. 12, pp. 2477–2486, 1999.

[70] M. Icardi, E. Gavi, D. L. Marchisio, A. A. Barresi, M. G. Olsen, R. O. Fox, and D. Lakehal, "Investigation of the flow field in a three-dimensional confined impinging jets reactor by means of microPIV and DNS," *Chem. Eng. J.*, vol. 166, no. 1, pp. 294–305, 2011.

[71] M. Icardi, E. Gavi, D. L. Marchisio, M. G. Olsen, R. O. Fox, and D. Lakehal, "Validation of LES predictions for turbulent flow in a confined impinging jets reactor," *Appl. Math. Model.*, vol. 35, no. 4, pp. 1591–1602, 2011.

[72] E. Gavi, D. L. Marchisio, and A. A. Barresi, "CFD modelling and scale-up of confined impinging jet reactors," *Chem. Eng. Sci.*, vol. 62, no. 8, pp. 2228–2241, 2007.

[73] . Makowski and J. Badyga, "Large Eddy Simulation of mixing effects on the course of parallel chemical reactions and comparison with k-epsilon modeling," *Chem. Eng. Process.: Process Intens.*, vol. 50, no. 10, pp. 1035–1040, 2011.

[74] F. Schwertfirm and M. Manhart, "A numerical approach for simulation of turbulent mixing and chemical reaction at high Schmidt numbers," in *Micro and Macro Mixing*. Springer, 2010, pp. 305–324.

[75] R. Santos, E. Erkoç, M. M. Dias, A. M. Teixeira, and J. C. B. Lopes, "Hydrodynamics of the mixing chamber in RIM: PIV flow-field characterization," *AIChE J.*, vol. 54, no. 5, pp. 1153–1163, 2008.

[76] R. J. Santos, A. M. Teixeira, E. Erkoç, M. Sultan, A. M. Karpinska, M. M. Dias, and J. C. B. Lopes, "Validation of a 2D CFD model for hydrodynamics' studies in CIJ mixers," *Int. J. Chem. React. Eng.*, vol. 8, no. 1, 2010.

[77] A. Soleymani, E. Kolehmainen, and I. Turunen, "Numerical and experimental investigations of liquid mixing in T-type micromixers," *Chem. Eng. J.*, vol. 135, pp. S219–S228, 2008.

[78] L. Metzger and M. Kind, "On the mixing in confined impinging jet mixers–Time scale analysis and scale-up using CFD coarse-graining methods," *Chem. Eng. Res. Des.*, vol. 109, pp. 464–476, 2016.

[79] L. Metzger and M. Kind, "Influence of mixing on particle formation of fast precipitation reactions–A new coarse graining method using CFD calculations as a "measuring" instrument," *Chem. Eng. Res. Des.*, 2016.

[80] J. Baldyga, J. R. Bourne, and S. J. Hearn, "Interaction between chemical reactions and mixing on various scales," *Chem. Eng. Sci.*, vol. 52, no. 4, pp. 457–466, 1997.

[81] J. Baldyga and J. R. Bourne, *Turbulent mixing and chemical reactions*. Wiley, 1999.

[82] A. A. Öncül, G. Janiga, and D. Thevenin, "Comparison of various micromixing approaches for computational fluid dynamics simulation of barium sulfate precipitation in tubular reactors," *Ind. Eng. Chem. Res.*, vol. 48, no. 2, pp. 999–1007, 2008.

[83] M. Harada, "Micro-mixing in a continuous flow reactor (coalescence and redispersion models)." *Memoirs of the Fac. of Eng., Kyoto University*, vol. 24, no. 4, 1962.

[84] P. Costa and C. Trevissoi, "Reactions with non-linear kinetics in partially segregated fluids," *Chem. Eng. Sci.*, vol. 27, no. 11, pp. 2041–2054, 1972.

[85] D. C. Haworth, "Progress in probability density function methods for turbulent reacting flows," *Prog. Energy Combust. Sci.*, vol. 36, no. 2, pp. 168–259, 2010.

[86] L. Falk and E. Schaer, "A PDF modelling of precipitation reactors," *Chem. Eng. Sci.*, vol. 56, no. 7, pp. 2445–2457, 2001.

[87] D. L. Marchisio, A. A. Barresi, and M. Garbero, "Nucleation, growth, and agglomeration in barium sulfate turbulent precipitation," *AIChE J.*, vol. 48, no. 9, pp. 2039–2050, 2002.

[88] L. Wang and R. O. Fox, "Comparison of micromixing models for CFD simulation of nanoparticle formation," *AIChE J.*, vol. 50, no. 9, pp. 2217–2232, 2004.

[89] J. Akroyd, A. J. Smith, L. R. McGlashan, and M. Kraft, "Numerical investigation of DQMOM-IEM as a turbulent reaction closure," *Chem. Eng. Sci.*, vol. 65, no. 6, pp. 1915–1924, 2010.

[90] K. Tsai and R. O. Fox, "PDF simulation of a turbulent series—parallel reaction in an axisymmetric reactor," *Chem. Eng. Sci.*, vol. 49, no. 24, pp. 5141–5158, 1994.

[91] Z. Li, J. Kessel, G. Grünewald, and M. Kind, "Coupled CFD-PBE simulation of nucleation in fluidized bed spray granulation," *Drying Technol.*, vol. 31, no. 15, pp. 1888–1896, 2013.

[92] K. M. Li, Z. and G. Grünewald, "Modeling the growth kinetics of fluidized-bed spray granulation," *Chem. Eng. & Tech.*, vol. 34, no. 7, pp. 1067–1075, 2011.

[93] M. Kucher, T. Beierlein, and M. Kind, "In situ WAXS synchrotron radiation study on particle formation of precipitated barium sulfate," *AIChE J.*, vol. 54, no. 5, pp. 1178–1188, 2008.

[94] J. R. Bourne and R. J. Demyanovich, "Rapid micromixing by the impingement of thin liquid sheets," *Ind. Eng. Chem. Res.*, vol. 28(6), p. 825, 1989.

[95] J. Baldyga and J. R. Bourne, "Simplification of micromixing calculations. I. Derivation and application of new model," *Chem. Eng. J.*, vol. 42, no. 2, pp. 83–92, 1989.

[96] H.-C. Schwarzer and W. Peukert, "Combined experimental/numerical study on the precipitation of nanoparticles," *AIChE J.*, vol. 50, no. 12, pp. 3234–3247, 2004.

[97] L. Falk and J.-M. Commenge, "Performance comparison of micromixers," *Chem. Eng. Sci.*, vol. 65, no. 1, pp. 405–411, 2010.

[98] K. Malecha, L. J. Golonka, J. Baldyga, M. Jasinska, and P. Sobieszuk, "Serpentine microfluidic mixer made in LTCC," *Sens. Actuators, B*, vol. 143, no. 1, pp. 400–413, 2009.

[99] J. Ottino, W. E. Ranz, and C. W. Macosko, "A lamellar model for analysis of liquid-liquid mixing," *Chem. Eng. Sci.*, vol. 34, no. 6, pp. 877–890, 1979.

[100] J.-M. Commenge and L. Falk, "Villermaux-Dushman protocol for experimental characterization of micromixers," *Chem. Eng. Process.*, vol. 50, no. 10, pp. 979–990, 2011.

[101] J. Baldyga and R. Pohorecki, "Turbulent micromixing in chemical reactors-a review," *Chem. Eng. J. and the Biochem. Eng. J.*, vol. 58, no. 2, pp. 183–195, 1995.

[102] J. Baldyga and J. R. Bourne, "A fluid mechanical approach to turbulent mixing and chemical reaction part II micromixing in the light of turbulence theory," *Chem. Eng. Commun.*, vol. 28, no. 4-6, pp. 243–258, 1984.

[103] D. L. Marchisio, L. Rivautella, and A. A. Barresi, "Design and scale-up of chemical reactors for nanoparticle precipitation," *AIChE J.*, vol. 52, no. 5, pp. 1877–1887, 2006.

[104] V. Skorych, M. Dosta, E.-U. Hartge, and S. Heinrich, "Dyssol–Modeling framework for the dynamic flowsheet simulation of solids processes," *CIT*, vol. 88, no. 9, pp. 1357–1357, 2016.

[105] A. Mersmann, "Calculation of interfacial tensions," *J. Cryst. Growth*, vol. 102, no. 4, pp. 841–847, 1990.

[106] M. Ailinger, "Bachelor thesis: Untersuchung zur Fällung schwerlöslicher Salze," *Karlsruhe Institute of Technology - Institute of Thermal Process Engineering*, 2015.

[107] R. Zauner and A. G. Jones, "On the influence of mixing on crystal precipitation processes—application of the segregated feed model," *Chem. Eng. Sci.*, vol. 57, no. 5, pp. 821–831, 2002.

[108] *DIN 28131 – Agitators and baffles for agitator vessels; types, terms and main dimensions*, Beuth. Std., 1992-09.

[109] W. Gösele and M. Kind, "Versuche zum Einfluß der Vermischung auf die Qualität eines kontinuierlich gefällten Produktes," *CIT*, vol. 63, no. 1, pp. 59–62, 1991.

[110] M. L. J. Van Leeuwen, O. S. L. Bruinsma, and G. M. Van Rosmalen, "Three-zone approach for precipitation of barium sulphate," *J. Cryst. Growth*, vol. 166, no. 1, pp. 1004–1008, 1996.

[111] A. H. Alexopoulos, D. Maggioris, and C. Kiparissides, "CFD analysis of turbulence non-homogeneity in mixing vessels: a two-compartment model," *Chem. Eng. Sci.*, vol. 57, no. 10, pp. 1735–1752, 2002.

[112] J. R. Bourne and S. Yu, "Investigation of micromixing in stirred tank reactors using parallel reactions," *Ind. Eng. Chem. Res.*, vol. 33, no. 1, pp. 41–55, 1994.

[113] E. Kougoulos, A. G. Jones, and M. Wood-Kaczmar, "CFD modelling of mixing and heat transfer in batch cooling crystallizers: Aiding the development of a hybrid predictive compartmental model," *Chem. Eng. Res. Des.*, vol. 83, no. 1, pp. 30–39, 2005.

[114] V. Kulikov, H. Briesen, and W. Marquardt, "A framework for the simulation of mass crystallization considering the effect of fluid dynamics," *Chem. Eng. Process.: Proc. Intens.*, vol. 45, no. 10, pp. 886–899, 2006.

[115] R. K. Geisler, A. Mersmann, and H. Voit, "Makro-und Mikromischen im Rührkessel," *CIT*, vol. 60, no. 12, pp. 947–955, 1988.

[116] V.-G. V. und Chemieingenieurwesen (Ed.). VDI Heat Atlas, *chap. N3*. Springer, Heidelberg/Dordrecht/London/New York, 2010.

[117] J. Baldyga, W. Podgorska, and R. Pohorecki, "Mixing-precipitation model with application to double feed semibatch precipitation," *Chem. Eng. Sci.*, vol. 50, no. 8, pp. 1281–1300, 1995.

[118] J. A. Denev, J. Fröhlich, and H. Bockhorn, "Evaluation of mixing and chemical reactions within a jet in crossflow by means of LES," in *Proc. European Combustion Meeting, April*, 2005, pp. 3–6.

[119] S. H. Smith and M. G. Mungal, "Mixing, structure and scaling of the jet in crossflow," *J. Fluid Mech.*, vol. 357, pp. 83–122, 1998.

[120] A. Fluent, "14.0 Theory Guide," *ANSYS inc*, 2011.

[121] H. A. Jakobsen, "Agitation and fluid mixing technology," in *Chem. React. Model.* Springer, 2014, pp. 809–881.

[122] M. Karimi, G. Akdogan, and S. M. Bradshaw, "Effects of different mesh schemes and turbulence models in CFD modelling of stirred tanks," *Physicochem. Probl. Mi.*, vol. 48, no. 2, p. 513e31, 2012.

[123] J. Baldyga, W. Orciuch *et al.*, "Double-feed semibatch precipitation effects of mixing," *Chem. Eng. Res. Des.*, vol. 85, no. 5, pp. 745–752, 2007.

[124] R. K. Geisler, "Fluiddynamik und Leistungseintrag in turbulent gerührten Suspensionen," Ph.D. dissertation, TU München, 1991.

[125] J. Baldyga and J. R. Bourne, "Interactions between mixing on various scales in stirred tank reactors," *Chem. Eng. Sci.*, vol. 47, no. 8, pp. 1839–1848, 1992.

[126] S. Motamedvaziri and P. M. Armenante, "Flow regimes and surface air entrainment in partially filled stirred vessels for different fill ratios," *Chem. Eng. Sci.*, vol. 81, pp. 231–250, 2012.

[127] J. R. Bourne, R. V. Gholap, and V. B. Rewatkar, "The influence of viscosity on the product distribution of fast parallel reactions," *Chem. Eng. J.*, vol. 58, no. 1, pp. 15–20, 1995.

[128] J. Baldyga, M. Henczka, and L. Makowski, "Effects of mixing on parallel chemical reactions in a continuous-flow stirred-tank reactor," *Chem. Eng. Res. Des.*, vol. 79, no. 8, pp. 895–900, 2001.

[129] P. Atkins and J. De Paula, "Atkins' physical chemistry," *Oxford, New York*, 2006.

[130] E. Samson, J. Marchand, and K. A. Snyder, "Calculation of ionic diffusion coefficients on the basis of migration test results," *Mater. Struct.*, vol. 36, no. 3, pp. 156–165, 2003.

[131] GESTIS, "Stoffdatenbank," *URL: http://www. hvbg. de,*, 2003.

[132] R. Budwig, "Refractive index matching methods for liquid flow investigations," *Exp. Fluids*, vol. 17, no. 5, pp. 350–355, 1994.

[133] Y. A. Hassan and E. E. Dominguez-Ontiveros, "Flow visualization in a pebble bed reactor experiment using PIV and refractive index matching techniques," *Nucl. Eng. Des.*, vol. 238, no. 11, pp. 3080–3085, 2008.

[134] T. L. Narrow, M. Yoda, and S. Abdel-Khalik, "A simple model for the refractive index of sodium iodide aqueous solutions," *Exp. Fluids*, vol. 28, no. 3, pp. 282–283, 2000.

[135] J. R. Warren and J. A. Gordon, "On the refractive indices of aqueous solutions of urea," *J. Phys. Chem.*, vol. 70, no. 1, pp. 297–300, 1966.

[136] J. Hála, "IUPAC-NIST solubility data series. 79. Alkali and alkaline earth metal pseudohalides," *J. Phys. Chem. Ref. Data*, vol. 33, no. 1, pp. 1–176, 2004.

[137] G. Jones and M. Dole, "The viscosity of aqueous solutions of strong electrolytes with special reference to barium chloride," *J. Am. Chem. Soc.*, vol. 51, no. 10, pp. 2950–2964, 1929.

[138] M. Bohnet, *Mechanische Verfahrenstechnik*. John Wiley & Sons, 2012.

[139] P. R. Wiederhold, *Water vapor measurement: methods and instrumentation*. CRC Press, 1997.

[140] L. L. Beranek and I. L. Ver, "Noise and vibration control engineering-principles and applications," *814 p., John Wiley & Sons, Inc.*, vol. 1, 1992.

[141] M. Cudina and J. Prezelj, "Detection of cavitation in situ operation of kinetic pumps: Effect of cavitation on the characteristic discrete frequency component," *Appl. Acoust.*, vol. 70, no. 9, pp. 1175–1182, 2009.

[142] G. K. Batchelor, *An introduction to fluid dynamics*. Cambridge university press, 2000.

[143] B. M. Fabuss, A. Korosi, and A. S. Hug, "Densities of binary and ternary aqueous solutions of NaCl, Na2SO4 and MgSO4 of sea waters, and sea water concentrates." *J. Chem. Eng. Data*, vol. 11, no. 3, pp. 325–331, 1966.

[144] A. L. Surdo, E. M. Alzola, and F. J. Millero, "The (p, V, T) properties of concentrated aqueous electrolytes I. Densities and apparent molar volumes of NaCl, Na2SO4, MgCl2, and MgSO4 solutions from 0.1 mol kg- 1 to saturation and from 273.15 to 323.15 K," *J. Chem. Thermodyn.*, vol. 14, no. 7, pp. 649–662, 1982.

[145] D. Puchalska and G. Atkinson, "Densities and apparent molal volumes of aqueous BaCl2 solutions from 15 to 140 °C and from 1 to 200 bar," *J. Chem. Eng. Data*, vol. 36, no. 4, pp. 449–452, 1991.

[146] J. Kestin, M. Sokolov, and W. A. Wakeham, "Viscosity of liquid water in the range -8 C to 150 C," *J. Phys. Chem. Ref. Data*, vol. 7, no. 3, pp. 941–948, 1978.

[147] M. Kaminsky, "The concentration and temperature dependence of the viscosity of aqueous solutions of strong electrolytes. III. KCl, K2SO4, MgCl2, BeSO4, and MgSO4 solutions," *Z. phys. Chem*, vol. 12, pp. 206–231, 1957.

[148] I. M. Abdulagatov, A. Zeinalova, and N. D. Azizov, "Viscosity of aqueous Na2SO4 solutions at temperatures from 298 to 573K and at pressures up to 40MPa," *Fluid Phase Equilib.*, vol. 227, no. 1, pp. 57–70, 2005.

[149] T. Isono, "Density, viscosity, and electrolytic conductivity of concentrated aqueous electrolyte solutions at several temperatures. Alkaline-earth chlorides, lanthanum chloride, sodium chloride, sodium nitrate, sodium bromide, potassium nitrate, potassium bromide, and cadmium nitrate," *J. Chem. Eng. Data*, vol. 29, no. 1, pp. 45–52, 1984.

[150] P. Wang, A. Anderko, and R. D. Young, "A speciation-based model for mixed-solvent electrolyte systems," *Fluid Phase Equilib.*, vol. 203, no. 1, pp. 141–176, 2002.

[151] M. M. Lencka, A. Anderko, S. J. Sanders, and R. D. Young, "Modeling viscosity of multicomponent electrolyte solutions," *Int. J. Thermophys.*, vol. 19, no. 2, pp. 367–378, 1998.

[152] J. B. Robinson, J. M. Strottmann, and E. Stellwagen, "Prediction of neutral salt elution profiles for affinity chromatography," *Proc. Natl. Acad. Sc.*, vol. 78, no. 4, pp. 2287–2291, 1981.

[153] J. J. Stickel and R. L. Powell, "Fluid mechanics and rheology of dense suspensions," *Annu. Rev. Fluid Mech.*, vol. 37, pp. 129–149, 2005.

[154] V. Yakhot, S. Orszag, S. A .and Thangam, T. B. Gatski, and C. G. Speziale, "Development of turbulence models for shear flows by a double expansion technique," *Phys. Fluids A : Fluid Dyn.*, vol. 4, no. 7, pp. 1510–1520, 1992.

[155] K. Ng, N. J. Fentiman, K. C. Lee, and M. Yianneskis, “Assessment of sliding mesh CFD predictions and LDA measurements of the flow in a tank stirred by a Rushton impeller,” *Chem. Eng. Res. Des.*, vol. 76, no. 6, pp. 737–747, 1998.

[156] M. A. J. Hartig, W. Peukert, N. Jacobsen, and A. Leuthold, “A model-based precipitation study of copper-based catalysts,” *AIChE Journal*, vol. 61, no. 7, pp. 2104–2116, 2015.

[157] R. G. Gordon, “Error bounds in equilibrium statistical mechanics,” *J. Math. Phys.*, vol. 9, no. 5, pp. 655–663, 1968.

8 Appendix

8.1 Methods and Materials

8.1.1 Reference potentials

Equations 8.1 to 8.3 show the exact expression for the reference potentials used in eq. 2.18, 2.19 and 2.20 with respect to the basic reference potential of the pure component ($\mu_{0i}(p,T)$). Detailed explanation can be found in Schaber [19].

$$\mu_{\infty,i} = \mu_{0,i}(p,T) + RT\, ln\, \gamma_{\infty,i} \qquad 8.1$$

$$\mu_{\infty,\tilde{c}_i} = \mu_{0,i}(p,T) + RT\, ln \frac{\gamma_{\infty,i}\tilde{c}_{0,i}}{\tilde{c}_\infty} \qquad 8.2$$

$$\mu_{\infty,\tilde{m}_i} = \mu_{0,i}(p,T) + RT\, ln\, \gamma_{\infty,i}\widetilde{M}_i\, \tilde{m}_{0,i} \qquad 8.3$$

8.1.2 Material data used for $BaSO_4$

Table 8.1: Material data used for $BaSO_4$

Parameter	*Value*
solid density $\rho^S/(kg/m^3)$	4480
solubility product $K_{sp}/\,(mol^2/l^2)$ [28]	$9.82 \cdot 10^{-11}$
molar mass $\widetilde{M}^S/(g/mol)$	233.34
interfacial energy $\gamma^{SL}/(J/m^2)$ [5]	0.1181

Table 8.2: Comparison of diffusion coefficients calculated with the Stokes-Einstein approach and those listed in literature at 25 °C

Diffusion coefficients	(m^2/s)
$D^L_{Ba^{2+},H_2O}$ [129]	$7.43 \cdot 10^{-10}$
$D^L_{SO_4^{2-},H_2O}$ [130]	$1.065 \cdot 10^{-9}$
$D^L_{Ba^{2+},H_2O}$ (Stokes-Einstein)	$6.063 \cdot 10^{-10}$
$D^L_{SO_4^{2-},H_2O}$ (Stokes-Einstein)	$6.464 \cdot 10^{-10}$

The diffusion coefficients according to Stokes-Einstein are calculated with eq. 8.4 with the hydrate radius of the ions $r_{H,Ba^{2+}} = 4.04 \cdot 10^{-10}\, m$, $r_{H,SO_4^{2-}} =$

$3.79 \cdot 10^{-10}\, m$, the temperature T= $298\, K$ and the viscosity $\eta^L =$ $0.891\, mPas$ [30].

$$D_{AB}^L = \frac{k_B \cdot T}{6 \cdot \pi \cdot \eta_L \cdot r_{H,AB}} \tag{8.4}$$

Mersmann [8] propose an apparent diffusion coefficient from the harmonic mean of the single-ion diffusion coefficients in the liquid phase, consonant with eq. 8.1. ($A = Ba^{2+}, B = SO_4^{2-}$), z_A and z_B represents the charge of the ions ($= 2$).

$$D_{AB}^L = \frac{(z_A + z_B) \cdot D_A^L D_B^L}{z_A \cdot D_A^L + z_B \cdot D_B^L} \tag{8.5}$$

Table 8.3: Mean diffusion coefficients for $BaSO_4$, following eq. 8.5.

Mean diffusion coefficient (eq. 8.5)	(m^2/s)
D_{AB}^L (literature)	$8.736 \cdot 10^{-10}$
D_{AB}^L (Stokes-Einstein)	$6.257 \cdot 10^{-10}$

Kucher et al. [21] propose a mean diffusion coefficient (using Stokes-Einstein) of $D_{BaSO_4}^L = 1.1 \cdot 10^{-9}\, m^2/s$, assuming $r_{h,\text{mean}} = (V_m)^{1/3}/\,2 = 2.21 \cdot 10^{-10}\, m$. Simulations have been applied but no big influence on the particle formation and the final particle distribution has been found for either the lowest or the highest diffusion coefficient value shown. The suggestion of Kucher et al. [21] is used, which enables the calculation of the diffusion coefficient from the molar volume V_m without additional information, for the sake of flexibility in the flowsheet simulation (chapter 1). Readers should keep in mind that this can lead to slight overestimations of the diffusion coefficient, since, for example, the additional expansion of the solvated ions by the hydrate shell is neglected.

Maximal adjustable supersaturations for the material system investigated

Natural limits for the maximal possible supersaturation adjustable are set by the maximal solubility of the reactants for the precipitation reaction regarded.

$$BaCl_2(aq) + Na_2SO_4(aq) \rightarrow BaSO_4 \downarrow + 2\, NaCl(aq)$$

Table 8.4 gives the solubility limits for the reactants selected at $20\ °C$ from literature. By inserting these limits into the Pitzer activity calculations, one can get Figure 8.1.

Table 8.4: Solubility of the reactants at $\vartheta = 20\ °C$

Reactant	Solubility limit / (mol/l)
$BaCl_2$	1.80072 [131]
Na_2SO_4	1.196856 [131]

The maximal feasible supersaturation varies depending on the momentum ratio and the free lattice ion ratio $\mathrm{R_c}$ chosen. This look-up chart can be of high interest for parameter studies. The maximal possible value is found to be $S_{a,\mathrm{nom,max}} = 1842$ at $\mathrm{R_c} = 1.562$ for a momentum ratio $\mathrm{M_R}$ of 1.

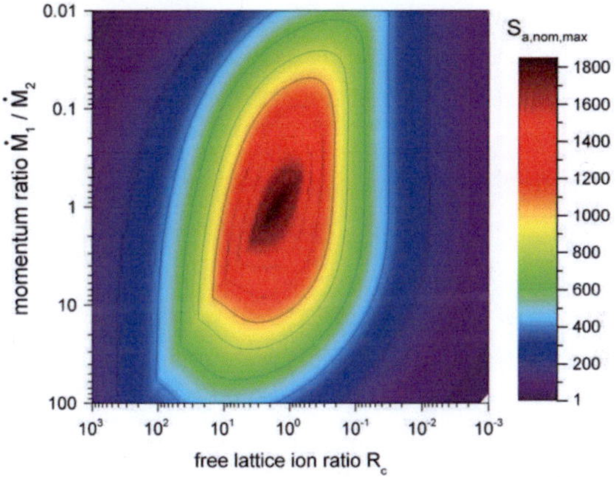

Figure 8.1: Maximal feasible supersaturation for $BaSO_4$ at $20\ °C$ under varying mixing conditions. Thereby, the solubility product of the aqueous reactant solution is the limiting factor.

8.2 Confined Impinging Jet Mixers

8.2.1 Refractive indices correction for LIF and PIV

The laser stimulation of the control volume causes fluorescence light emission (LIF) and light emission by the tracer particles (PIV). The light emitted is diffracted by the cylindrical mixing geometry. This leads to imperfect intensity information concerning the LIF measurements and, furthermore, to incorrect local information for the tracer particles in the near wall region during PIV investigation. These undesirable effects can be suppressed with a refractive index correction between fluid and PMMA-mixing geometry.

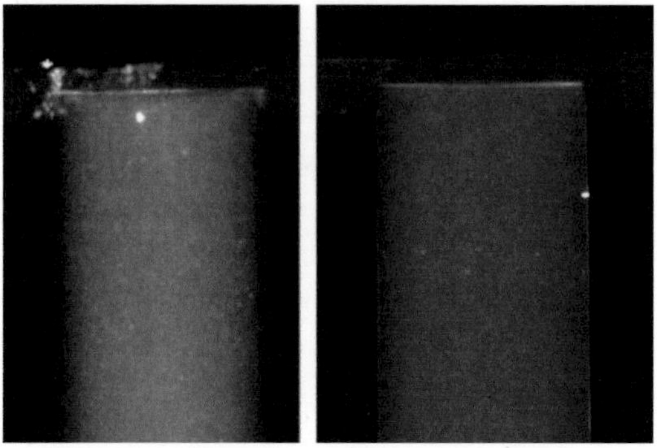

Figure 8.2: PIV image with Rhodamine B addition of the PMMA-T-Mixer without (left) and with (right) refractive index correction.

The challenge for this correction is to conform the refractive index of the liquid phase ($n_{ri,H_2O} = 1.33$) with appropriate additives to that of PMMA ($n_{ri,\mathrm{PMMA}} = 1.49$ [132]) in due consideration of keeping the fluid properties almost the same. Similar viscosity and density as well as Newtonian behavior are basic requirements for the correction. Certain literature-based recommendations were tested on their applicability. Budwig [132] and Hassan and Dominguez-Ontiveros [133] give a wide overview of possible additives. Icardi et al. [70] used a urea water mixture. Budwig [132], Hassan and Dominguez-Ontiveros [133] and Narrow et al. [134] mention NaI and NH_4SCN as candidates with high potential. Important properties regarding the corrected aqueous solution are listed due to poor information in the literature.

Refractive indices data are generated with a refractometer (*Anton Paar, Abbemat*) at the wavelength of the sodium vapor lamp ($\lambda = 589.3\,nm$). Deviances caused by wavelengths are neglected due to the fact that only a small change in wavelength occurs in the $NdYAg$ laser application ($\lambda = 532\,nm$)[132]. Density data are achieved with a densitometer (*Anton Paar,*

DMA 5000), and viscosity data for the aqueous solution are measured with a cone and plate geometry ($d_{\text{cone}} = 60\, mm$) with a rotational rheometer in the shear range from $\dot{\gamma} = 1 - 500\; 1/s$ (Anton Paar, MCP 130).

Figure 8.3, left and right, show the dependency of the refractive index of the aqueous solution in the presence of different mass fractions of solute salt. The dashed line at 1.49 indicates the target value of *PMMA*.

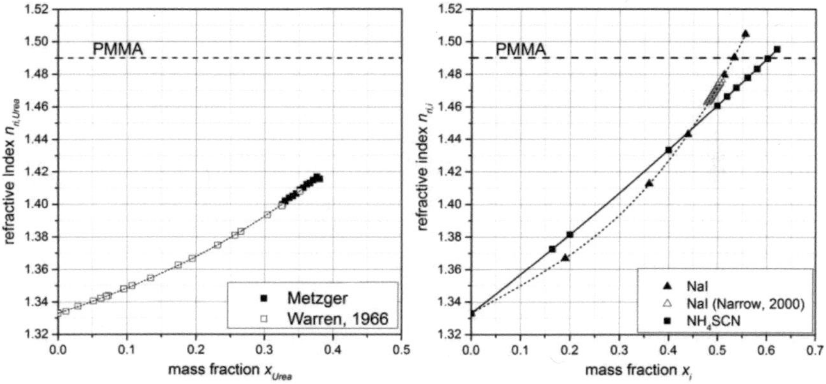

Figure 8.3: Left: Refractive index for urea [135]; right: NaI and NH_4SCN depending on the mass fraction for the salt regarded at 20 °C [134].

The measurements agree well with the literature. A polynomial approach describes the data for all three aqueous mixtures.

$$n_{ri,i}(20°C) = A + B \cdot x_i + C \cdot x_i^2 + D \cdot x_i^3 \qquad 8.6$$

Table 8.5: Polynomial coefficients for the refractive index correlations

Additive	A	B	C	D
Urea	1.33292	0.14272	0.10785	0.27092
NaI	1.33312	0.16751	-0.04638	0.53752
NH_4SCN	1.33296	0.23349	0.0457	-0.00105

The data covers the entire range from pure water up to the solubility limit of the relevant salt. One can see that urea is not able to lift the refractive index of the solution up to that of PMMA. When reaching the solubility limit, a maximal refractive index of 1.41 is adjusted. Both NaI and NH_4SCN fulfill this requirement. The addition of salt to the solution leads to an increase in viscosity that, however, remains Newtonian. Experiments are realized with the favorable ammonium thiocyanate due to the higher costs of sodium iodide. It should be mentioned that another advantageous aspect of the system water-NH_4SCN for refractive correction is the equality between the density of polyamide tracer particles and the density of the solution that eliminates unwanted buoyancy effects ($\rho_{\mathrm{PA}}^{S} = \rho^{L}$=1.14 kg/l).

Table 8.6 gives a short overview. Figure 8.4 shows viscosity and density data determined experimentally for NH_4SCN at $20\,°C$. Although a clear trend exists for a rising viscosity with the increasing mass fraction of salt, common mathematical correlations for viscosity data of electrolytes in aqueous solutions, such as the Jones Dole fit or some extension of the Jones Dole Fit, do not fit the data well. The wide measurement range ($\tilde{c}^{*}_{NH_4SCN} = 21.31\ mol/l$ [136]) leads to an excess of the area of validity for this kind of fit (Jones Dole $1\ mol/l$ [137]) due to a very high solubility limit. It should be mentioned that another advantageous aspect of the system water-NH_4SCN for refractive correction is the equality between the density of polyamide tracer particles and the density of the solution that eliminates unwanted buoyancy effects ($\rho_{\mathrm{PA}}^{S} = \rho^{L}$=1.14 kg/l).

Table 8.6: Polynomial overview of the characteristic data of prospective ionic solutions for refractive index correction

Additive	possible range of refractive index	Relative viscosity for $n_{ri,\mathrm{solution}} = n_{ri,\mathrm{PMMA}}$	Cost	Fulfils the requirements
Urea	1.33-1,41	-	+	-
NaI	1.33-1.51	2.55	+++	+
NH_4SCN	1.33-1.50	2.11	++	+

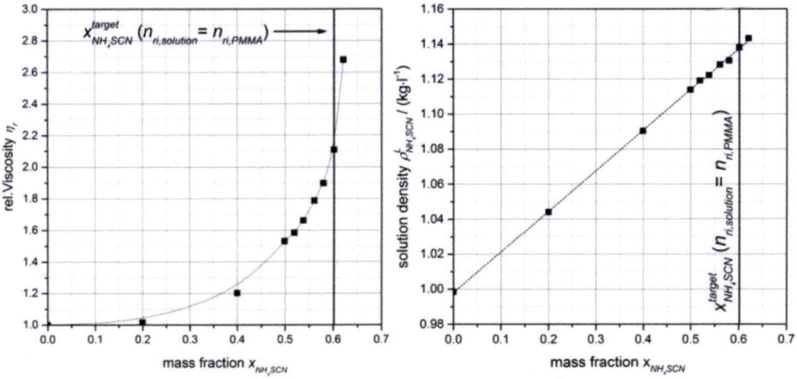

Figure 8.4: Relative dynamic viscosity $\eta_r = \eta/\eta_{H2O}$ of aqueous NH_4SCN solutions, dashed line as trend line (left), density of aqueous NH_4SCN solutions (right).

8.2.2 Flow regimes in CIJMs – CFD findings

Figure 8.5 presents snapshots of transient 3D DES SST k-ω simulations at different flow rates.

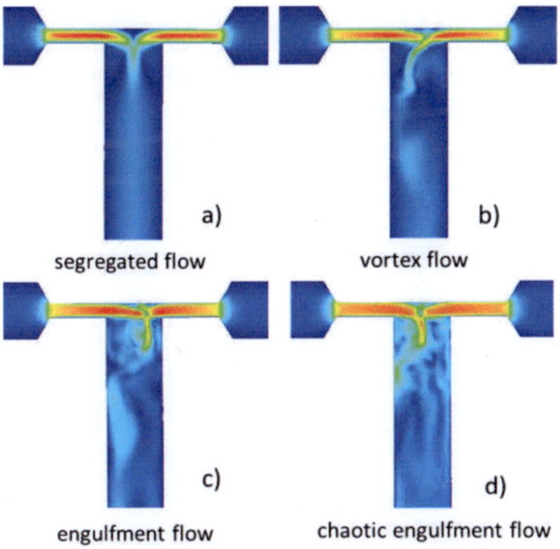

Figure 8.5: Mean velocities in the T-CIJM at different flow rates; a) $\mathrm{Re}_{\mathrm{mix}} = 20$, b) $\mathrm{Re}_{\mathrm{mix}} = 100$, c) $\mathrm{Re}_{\mathrm{mix}} = 400$ and d) $\mathrm{Re}_{\mathrm{mix}} = 3000$.

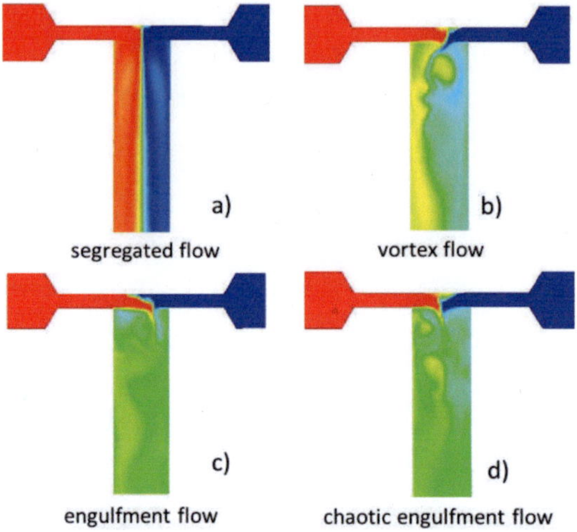

Figure 8.6: Species distribution (red and blue) in the mixer at different flow rates; a) $\mathrm{Re}_{\mathrm{mix}} = 20$*, b)* $\mathrm{Re}_{\mathrm{mix}} = 100$*, c)* $\mathrm{Re}_{\mathrm{mix}} = 400$ *and d)* $\mathrm{Re}_{\mathrm{mix}} = 3000$.

Thereby, flow regimes found experimentally and confirmed by literature can be well reproduced by the CFD simulations (classified in Table 3.3). Additionally, the species distribution of two aqueous liquids (red and blue) is given to demonstrate the different mixing intensities in the specific flow regimes. Figure 8.5 and Figure 8.6 demonstrate clearly how flow intensities and predominant flow regimes influence the mixing characteristics in a CIJM.

8.2.3 Additional STAR NM results

Step 1: Supersaturation buildup

The STAR NM data presented in section 3.4.3 outline only selected data. A more comprehensive database on the CFD "measurements" is given in this chapter. The spatially and temporally averaged supersaturation along a spatial coordinate (mixer length L_{mix}) is measured in step 1. Figure 8.7 shows the influence of the $\mathrm{Re}_{\mathrm{mix}}$ on the supersaturation buildup ($S_{a,\mathrm{nom}} = 1000$, $\mathrm{R_c} = 1$ are held constant).

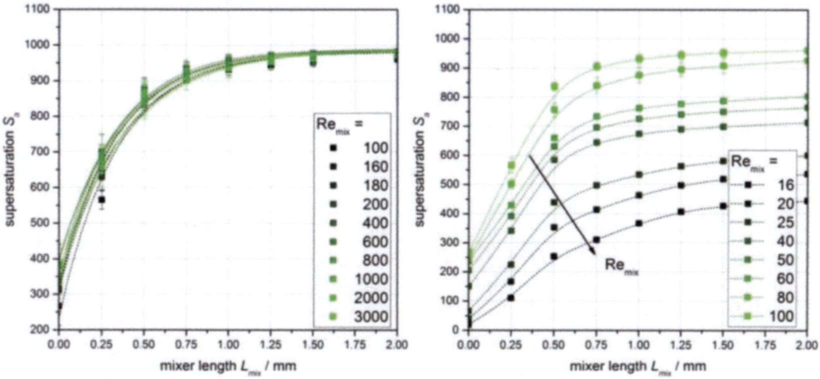

Figure 8.7: Left: Influence of the Re$_{mix}$ *number on the supersaturation buildup in the CIJM investigated (*$S_{a,\mathrm{nom}} = 1000$*,* $R_c = 1$*), right: Influence of the free lattice ion ratio on the supersaturation buildup.*

It can be shown that the supersaturation buildup for $\mathrm{Re}_{\mathrm{mix}} > 100$ (engulfment flow and chaotic engulfment flow) follows almost the same exponential curve. The mixing intensity decreased dramatically in the laminar regime $\mathrm{Re}_{\mathrm{mix}} \leq 100$, which also affects the supersaturation buildup (Figure 8.7, right). Figure 8.8 shows the influence of the free lattice ion ratio $\mathrm{R_c}$ adjusted. It can be observed for $\mathrm{R_c} > 1$ (Figure 8.8, left) and for $\mathrm{R_c} < 1$ (Figure 8.8, right) that supersaturation buildup is faster than in the stoichiometric case.

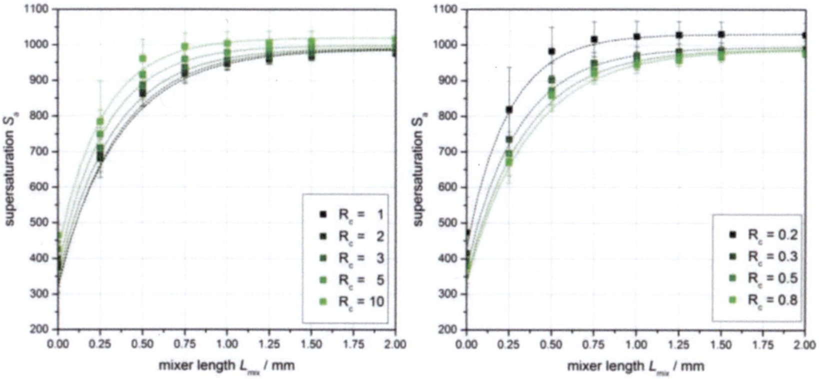

Figure 8.8: STAR NM data with exponential fit: left: Influence of $\mathrm{R_c} \geq 1$ *at* $\mathrm{Re}_{\mathrm{mix}} = 1000$ *and* $S_{a,\mathrm{nom}} = 1000$*, right: Influence of* $\mathrm{R_c} \leq 1$*.*

This behavior may also be interesting when comparing the mean particle sizes precipitated at different stoichiometric conditions, since the turbulent Damkoehler number is then also shifted slightly.

Extensive correlations of the data plotted in Figure 8.7 and Figure 8.8 can be found in the Bachelor thesis of M. Winkler, which has been carried out in conjunction with this work (see "List of publications").

Step 2: Residence time measurement

Figure 8.9 shows additional mean residence time data measured for varying flow intensities. All data measured can be well correlated with allometric fit ($L_{\mathrm{mix}} = a \cdot \tau_i^b$).

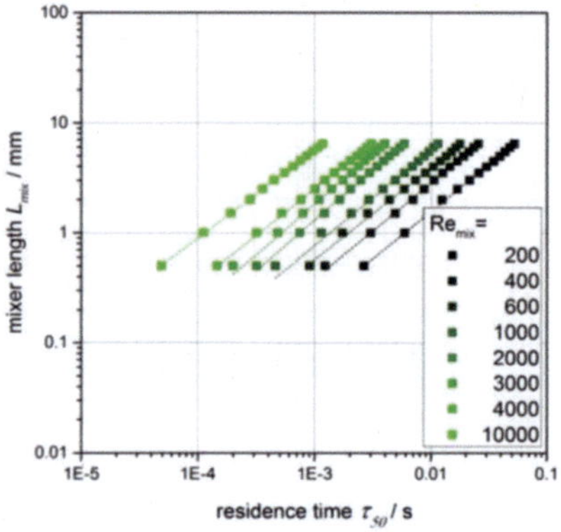

Figure 8.9: Residence time measurements in the T-CIJM shown exemplarily for $\mathrm{Re}_{\mathrm{mix}} \geq 200$.

All residence time data measured are correlated in Table 8.7. The mean residence time $\tau_{50,0}$ which is of special interest for the investigation made is resolved in more detail for the individual flow regimes.

Table 8.7: Empirical coefficients for the three residence time nodes $(\tau_{50}, \tau_{10}, \tau_{90})$ *to correlate the STAR NM results measured consonant with the relation.* $z_{\mathrm{mix}} = a_2 \cdot \tau_i^b$.

$\tau_i = \tau_{50}$	Regimes	Correlation parameters	Values
laminar	$10 \leq \mathrm{Re}_{\mathrm{mix}} \leq 100$	$a_{2.50}$	$6.3939 + 0.06662 \cdot \mathrm{Re}_{\mathrm{mix}}{}^{1.3571}$
		b_{50}	0.782
transition	$100 \leq \mathrm{Re}_{\mathrm{mix}} \leq 1500$	$a_{2.50}$	$0.5419 \cdot \mathrm{Re}_{\mathrm{mix}}{}^{0.9409}$
		b_{50}	0.86965

turbulent	$Re_{mix} \geq 1500$	$a_{2.50}$	$23.484 \cdot Re_{mix}^{0.44579}$
		b_{50}	0.86965
$\tau_i = \tau_{10}$	overall range	$a_{2.50}$	$2.4 \cdot Re_{mix}^{0.6085}$
		b_{50}	$1.1695 \cdot Re_{mix}^{-0.15831}$
$\tau_i = \tau_{90}$	overall range	$a_{2.50}$	$0.167 \cdot Re_{mix}^{1.241}$
		b_{50}	$1.1193 \cdot Re_{mix}^{0.02216}$

Correlation between Re_{mix} and mean energy dissipation in V_{mix}

Figure 8.10 gives a correlation between the Re_{mix} number adjusted in the CIJM and the mean energy dissipation $\bar{\varepsilon}$ "CFD-measured" in the mixing volume specified in section 3.4.3.7.

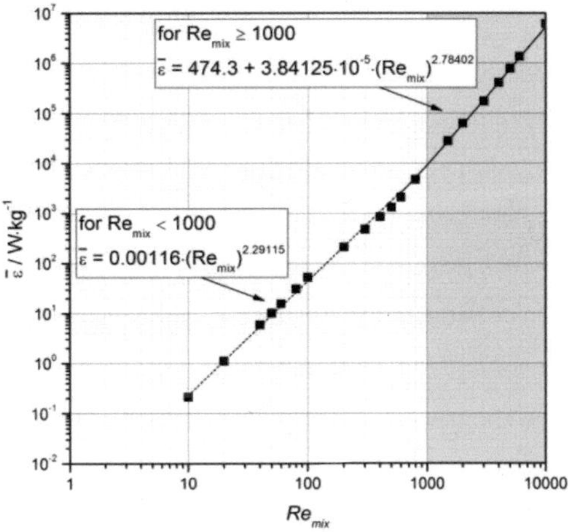

Figure 8.10: Correlation of Re_{mix} adjusted and the mean energy dissipation measured in V_{mix} in CFD.

Thereby, two ranges are specified: The turbulent range for $Re_{mix} > 1000$ and a range for lower Re_{mix} numbers. Thus, a relationship between the two flow parameters for the CIJM investigated can be created.

8.2.3.1 Axial dispersion in CIJMs

The measurements of the residence time distributions allow further considerations. Thus, the axial dispersion in CIJMs is investigated. According to Bohnet [138], the Bodenstein number Bo as a dimensionless measure of the dispersion in a system can be calculated with eq. 8.7 and is defined as eq. 8.8. The standard deviation of the residence time is calculated with $\sigma_\tau^2 = 0.5 \cdot (\tau_{10,0} - \tau_{50,0})^2 + (\tau_{90,0} - \tau_{50,0})^2$ using the residence time node information measured at the mixer outlet.

$$\sigma_\tau^2 = \frac{2}{\mathrm{Bo}} + \frac{8}{\mathrm{Bo}^2} \tag{8.7}$$

$$\mathrm{Bo} = \frac{\bar{u} \cdot L_{\mathrm{mix}}}{D_{\mathrm{ax}}} \tag{8.8}$$

Figure 8.11, left, shows Bo numbers calculated for the T-CIJM. An almost linear dependency on the $\mathrm{Re}_{\mathrm{mix}}$ number is observed. Increasing flow rates lead increasingly to an ideal plug flow behavior. This can also be stated regarding the axial dispersion coefficient in Figure 8.11, right, which decreases for a high $\mathrm{Re}_{\mathrm{mix}}$ number on values below the diffusivity of the single ions ($\approx 10^{-9}\ m^2/s$). These findings coincide well with experimental precipitation results, where particle sizes in the turbulent regime show a narrow, monomodal distribution.

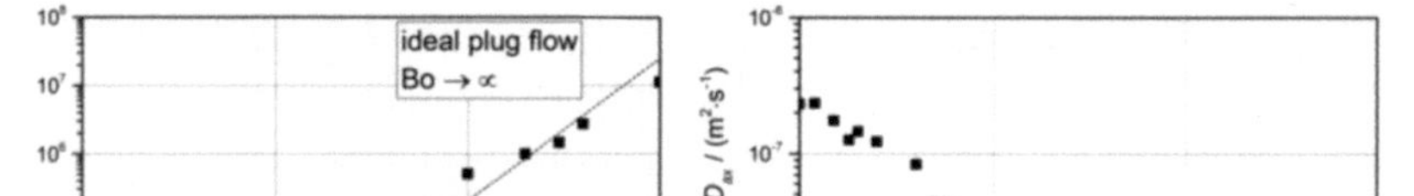

Figure 8.11: Left: Bodenstein numbers Bo *calculated to describe the axial dispersion of fluid in the CIJM for different flow rates, measured at the outlet of the mixer. Right: Axial dispersion coefficients* D_{ax} *calculated with eq. 8.8 at the mixer outlet (*$L_{\mathrm{mix}} = 6.5\ mm$*).*

8.2.3.2 Vortex frequencies

The STAR NM data measured, for example, in Figure 3.31, right, give the ability to predict characteristic vortex frequencies f_i from CFD. The frequency spectrum of the data measured for different $\mathrm{Re}_{\mathrm{mix}}$ numbers (see Figure 8.12, left) can be obtained by Fourier transformation. It can be shown that the maximal amplitude of the data measured shifts to higher frequencies for higher flow rates. Whereas a peak maximum at $\sim 40\ Hz$ for $\mathrm{Re}_{\mathrm{mix}} = 500$ can be identified, the maximum for $\mathrm{Re}_{\mathrm{mix}} = 6000$ is observed at $\sim 155\ Hz$. The Strouhal number Sr, defined consonant with eq. 8.9, is one possibility of accounting with a dimensionless number for the frequency of vortex shedding. The Sr numbers in Figure 8.12, right, are calculated by extracting the frequency of the highest amplitude. Although thee CFD "measured" frequencies have not yet been validated by experiments, it can be stated that the Strouhal numbers achieved cope well with those observed for other applications or natural phenomena. Values measured for a Kármán vortex street, for example, vary around 0.2.

$$\mathrm{Sr} = \frac{f_{\mathrm{max}} \cdot d_{\mathrm{mix}}}{\bar{u}_{\mathrm{mix}}} \qquad 8.9$$

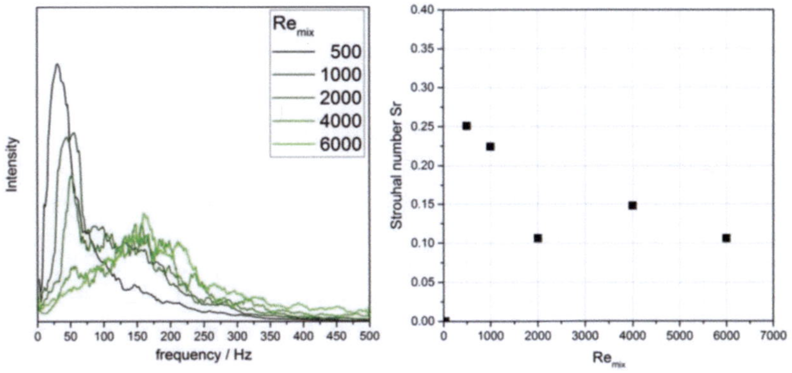

Figure 8.12: Left: Frequencies extracted by Fourier transformation from the STAR NM measurements for the T-CIJM. Right: Strouhal numbers calculated for the T-mixer.

8.2.4 Cavitation phenomena in CIJMs

If flow rates in both CIJMs investigated exceed a critical value $\mathrm{Re}_{\mathrm{mix,crit}}$, cavitation phenomena can be observed that lead to a transition of the flow regime. As discussed in section 3.3.1, the mixing structures of the impinging jets discharge their turbulent energy by forming distinctive vortices at the lower end. The onset of vortex cavitation is observed directly in the core axis of the vortices. In the process, the radial acceleration of the fluid lamella due to the vortex leads to a local reduction of static pressure below the vapor pressure of water ($p_{H_2O}^{V} = 31.7\ mbar$ at $298\,K$ [139]). Water suddenly vaporizes and forms small vapor bubbles. When these vapor bubbles leave the vortex core region, they implode, because of the higher ambient pressure of the surrounding bulk phase. A further increase in the flow rate leads to an intensification of the bubble formation and an increasing local two-phase flow on the first $\sim 5\ mm$ of the mixing zone. Vortex cavitation changes into cloud cavitation, where the mixer wall, as a heterogeneous surface, is also a source of cavitation bubbles. The LIF frames, as shown in Figure 3.6, illustrate operating sequences in the top of the mixing zone for the Y-mixer (top) and the T-mixer (bottom) at different $\mathrm{Re}_{\mathrm{mix}}$ numbers from 1600 to 5800.

A contact microphone was used for a sensitive detection of the initial cavitation point and for a better qualitative description of the cavitation phenomena. The implosion of vapor bubbles generates clearly indictable noise in the audible frequency range. Twenty s time-averaged frequency spectra of noise in the range from 0 to $22000\ Hz$ for specific flow rates are analyzed. The noise level $\mathcal{L}_p$ measured in *dB* is normalized by the lowest value captured in both mixers to get a better comparability for both devices.

$$\mathcal{L}_{p,N} = \frac{log\left(\frac{p_1^{\ 2}}{p_{\mathrm{ref}}^{\ 2}}\right)[dB]}{log\left(\frac{p_{\mathrm{min}}^{\ 2}}{p_{\mathrm{ref}}^{\ 2}}\right)[dB]} \qquad 8.10$$

$p_{\mathrm{ref}} = 2 \cdot 10^5\ Pa$ is the reference value for the airborne sound [140].

Only one specific peak in the noise frequency spectra for each mixing device is found to increase in the non-cavitating flow regime. Thus, this noise frequency was assigned to be the one of the transient mixing structures whose intensity of oscillation from one to the other mixer wall increases with increasing flow rate (Figure 8.13, left). The dashed lines mark the specific flow rates that are exemplarily presented in Figure 3.6.

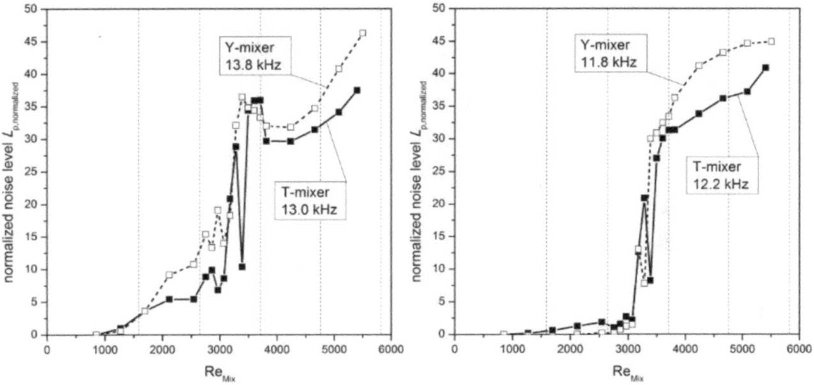

Figure 8.13: Characteristic cavitation frequencies (left) and characteristic vortex frequency (right). Dashed lines allow an attribution to Figure 3.6.

All frequencies other than these characteristic ones stay at a constant low noise level and increase suddenly with a factor of more than 40 on the log-scale of noise pressure level with the incipient vortex cavitation (Figure 8.13, right). Therefore, a clear transition from a non-cavitating to a cavitating regime can be observed at $\mathrm{Re}_{\mathrm{mix,crit}} \cong 3100$ for both mixers. This Re number is found to be the onset of the first cavitation bubbles in both mixers. It follows a transition region up to $\mathrm{Re}_{\mathrm{mix}} = 3400$ to build up continuous vortex cavitation. In contrast to the cavitation frequency chosen, the vortex cavitation frequency defined increases steadily for both mixers until $\mathrm{Re}_{\mathrm{mix,crit}}$ is reached. The following clear increase in the noise level is accompanied by a local minimum at $\mathrm{Re}_{\mathrm{mix}} = 4200$. Cudina and Prezelj [141] observed an early damping of noise in the system induced through the air bubbles after the initiation of cavitation. Subsequently, a further increase of the flow rate at $\mathrm{Re}_{\mathrm{mix}} = 4500$ follows a constant growth of the noise level once again.

Considering the hypothesis that CFD can also forecast this cavitation effect using the DES-SST-k-ω turbulence model, calculations at $\mathrm{Re}_{\mathrm{mix}} >$

$\mathrm{Re}_{\mathrm{mix,crit}} = 3700$ are carried out. Figure 8.14 shows a snapshot of the simulation.

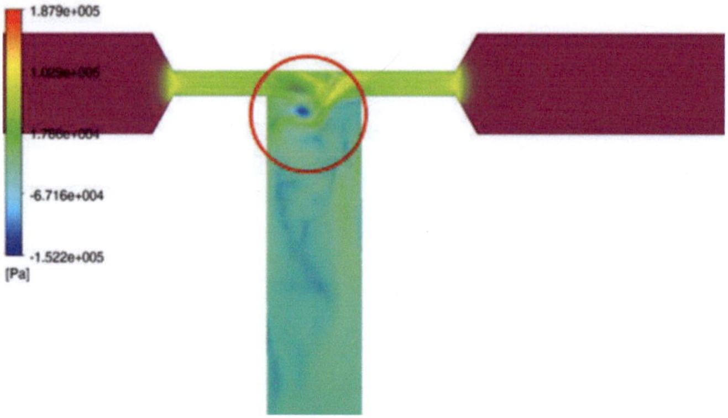

Figure 8.14: Superposed velocity field and static pressure in the T-mixer at $\mathrm{Re}_{\mathrm{mix}} > \mathrm{Re}_{\mathrm{mix,crit}} = 3700$*. The blue region in the mixer head indicates where the pressure is getting negative.*

One can see the pressure sink with a negative pressure value clearly straight behind the eddy shedding area. The transport equations for incompressible fluids are solved without pressure limits. A negative pressure that occurs in an incompressible fluid is to be interpreted as a driving force for cavitation [142]. Thus, the vortex core region serves as a birthplace of cavitation bubbles. A further increase of flow rates in the CFD simulation showed certain other low pressure areas that generate the driving force for vapor bubbles.

8.2.5 Introducing a density and rheology model for $BaSO_4$ suitable for the application within CFD

Appropriate models to consider the densities and viscosities of the ionic reactant solution and of the reacting solution in the mixing chamber of the CIJM correctly must be implemented within CFD. Since these quantities are an essential part within the Navier-Stokes equations, realistic models should be beneficial to achieving an accurate numeric solution. The effective viscosity of the ionic liquid–solids mixture, for instance, can reach twice the viscosity of pure water easily.

Measured density data of $BaCl_2$-water and Na_2SO_4-water at $\vartheta = 25\,°C$ are plotted in Figure 8.15, left, and are fitted with a first-order approximation (see Table 8.8). Experimental fits from literature (Na_2SO_4-water [143, 144]; $BaCl_2$-water [145]) at the temperature regarded are added for comparison.

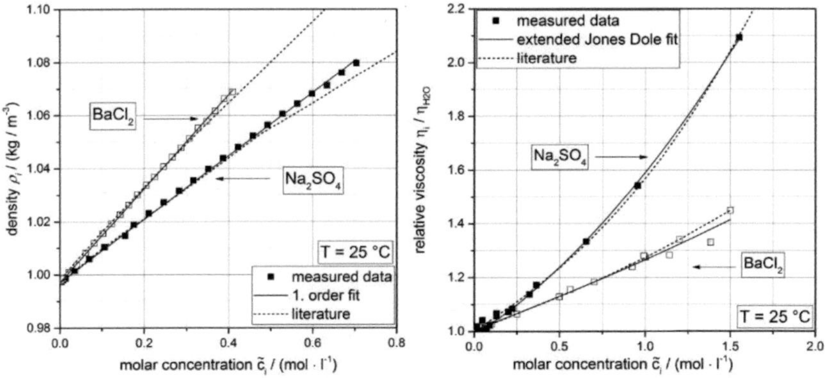

Figure 8.15: Left: Measured densities of the reactant solutions $BaCl_2$-water and Na_2SO_4-water. The dotted line represents the first-order approximation of the data (Table 8.8). Right: Reactant viscosities depending on the molar concentration of solvated salt i at 25 °C. The dashed lines are Kaminsky fits consonant with eq. 8.11.

The mixing of two feed streams and the disperse phase ($\rho^S_{BaSO_4} = 4480\ kg/m^3$), if it exists, is calculated with the volume weighted mixing law of the single phases involved. A small error due to the consumption of Ba^{2+} and SO_4^{2-} ions, which are integrated into the crystal structure and, therefore, no longer contribute to the density of the liquid phase, is tolerated, since the contribution of solids to the overall density is significantly higher.

Table 8.8: Linear fit parameter for the reactant density measured at 25 °C [146]

$\rho_i = m_1 \cdot \tilde{c}_i + b_0$	m_1	b_0 (density water at 25 °C)
$BaCl_2$	0.17676	0.99707
Na_2SO_4	0.11937	0.99707

The Newtonian viscosity behavior of the reactant solution depending on the molar concentration of solvated ions (see Figure 8.15, right) can be described well with an extended Jones Dole fit (Kaminsky) according to eq. 8.11 [147,

148]. Since the Jones Dole approach itself is only valid for $I_{\tilde{c}} < 0.1 - 0.2\,mol/l$, an additional fit, parameters D, has been added. Data from literature, Isono [149] for $BaCl_2$ and Abdulagatov et al. [148] for Na_2SO_4, are supplemented (dashed line) for comparison. The parameter A in eq. 8.11 was shown to be a function of solvent properties [150] and long-range electrostatic interactions [151], which are not operative at concentrations in excess of $20\,mM$ [152], while the viscosity parameter B can be related directly to the property of ions.

$$\frac{\eta_{\mathrm{Bulk},i}}{\eta_{H2O}} = 1 + A\tilde{c}_i^{\,1/2} + B\tilde{c}_i + D\tilde{c}_i^{\,2} \qquad 8.11$$

with $\eta_{H2O}(25\,°C) = 0.8903\,mPas$ [146].

Table 8.9: Extended Jones Dole parameter A, B and D (Kaminsky fit [147, 148]) at 25 °C for the reactant solutions $BaCl_2$-water and Na_2SO_4-water

Solvated salt i	A	B	D
$BaCl_2$	-0.01182	0.28123	0
Na_2SO_4	-0.05932	0.47136	0.17899

The Kaminsky fits, valid for the pure reactant solutions, do not work by themselves, since an adequate mixing law for the viscosity in every grid cell at every possible reactant concentration ratio is needed within the CFD calculations carried out in section 3.4.2. Therefore, a simplified mixing approach is introduced (see Figure 8.16, left). This approach only takes the viscosity parameter B that is related directly to the ion concentration into account. Higher-order contributions which could be considered with the unknown parameter D are neglected. The bulk viscosity of the aqueous solution containing $Ba^{2+}, Na^{+}, SO_4^{2-}$ and Cl^{-} ions is calculated according to eq. 8.12, wherein every type of ion has a special contribution to the viscosity depending on its single-ion Jones Dole B_i parameter.

$$\eta_{\mathrm{Bulk}} = \eta_{H_2O} \cdot \left(1 + \sum_{i=1}^{k} B_i \cdot \tilde{c}_i\right) \qquad 8.12$$

Table 8.10: Jones Dole parameters B_i

Ion i	B_i [152]
Ba^{2+}	0.22
Na^{+}	0.086
SO_4^{2-}	0.208
Cl^{-}	−0.007

The influence of solids on the bulk viscosity η_{Bulk} is measured (*Anton Paar MCR-Series*, cone-plate system) for different bulk fluid mixtures. This is established with different free lattice ion ratios $\mathrm{R_c}$, adjusted by the reactant concentrations, and is illustrated in Figure 8.16, right. Thereby, the relative (Newtonian) viscosity increase $\eta_{\mathrm{Susp}} / \eta_{\mathrm{Bulk}}$ is plotted against the solid volume fraction; wherein, η_{Bulk} for the different $\mathrm{R_c}$ ratios is calculated using eq. 8.12. A simple Einstein approach is not an adequate model to consider the contribution of the nano- and microparticles to viscosity observed in the ionic solution (see Figure 8.16, right), since viscosity increases significantly stronger in such substance systems than predicted by the simple hard sphere models. Therefore, an Eiler fit (eq. 8.13), which is a common empirical approach [153], has been adjusted to the data.

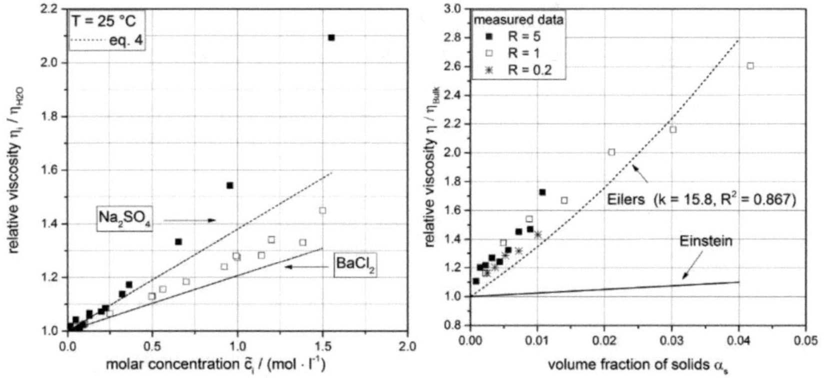

Figure 8.16: Left: Simplified model (eq. 8.12) to calculate the bulk viscosity of the ionic mixture using only the single-ion Jones Dole B_i coefficients, which can also be used in the reaction zone. Right: Influence of the solid volume fraction on the viscosity of the suspension. Thereby, differently composed ionic bulk solutions, indicated by the differing free ion ratio $\mathrm{R_c}$, are investigated.

Influences of the particle sizes ($\mathrm{R_c} = 5$: $d_p \triangleq nm$-particles → nanoparticles stabilized electrostatically, $\mathrm{R_c} = 0.2$ and $\mathrm{R_c} = 1 \triangleq \mu m$-particles → aggregate of nanoparticles [21]) on the viscosity have not been found. The measurements show that Newtonian behavior can be assumed for volume fractions below $\alpha_s < 0.01$ ($\triangleq S_a \approx 1400, \mathrm{R_c} = 5$).

For higher volume fractions, shear thinning should be taken into account, which will not be considered in this contribution.

$$\eta_{Susp} = \eta_{Bulk}\left(1 + \frac{k_1 \cdot \alpha_S}{1 - \alpha_S/\alpha_{S,m}}\right)^2 \qquad 8.13$$

Equation 8.13 contains the volume fraction of solids α_S, fit parameter $k_1 = 15.8$ and $\alpha_{S,m} = 0.68$ (packing density of a body-centered cubic elementary cell).

8.3 Dyssol – a simulation framework system based on C++

Figure 8.17 shows a screenshot of the simulation user-interface (*Dyssol*) used in this work (section 1 and 5.7).

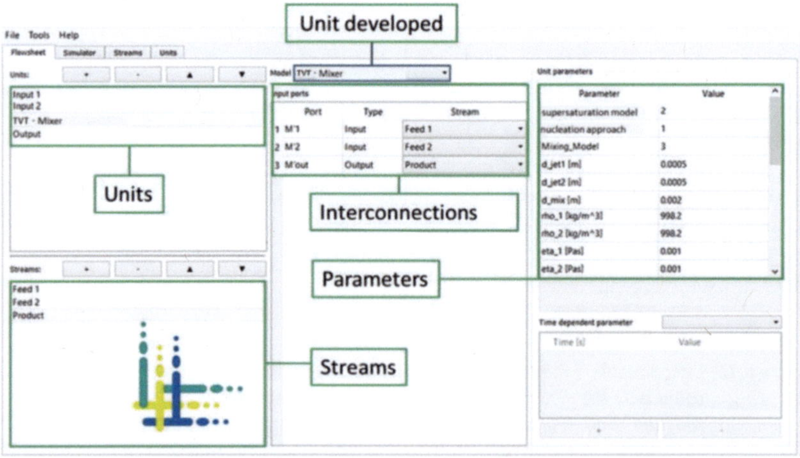

Figure 8.17: Screenshot of the simulation interface of the Dyssol software for CIJMs.

Thereby, units (CIJM and semi-batch ST reactor), input and output ports, constant and time-dependent parameters are specified in the process developed.

8.4 Stirred-tank reactors

This section provides additional results and material concerning the ST reactors investigated.

8.4.1 Influence of the feed pipe diameter

Figure 8.18 shows the influence of the internal feed pipe diameter $d_{i,\text{feed}}$ on the volumetric PSD at the end of the semi-batch process. Results are shown exemplarily for $n = 100\ rpm$ and $\dot{V}_{\text{feed}} = 100\ ml/min$. Larger particles and a stronger broadened distribution can be identified for an increasing diameter. The data measured confirm that the entrance impulse and, moreover, the dispersion of the reactants entering the ST have a significant influence on the emerging supersaturation field and the PSD formed.

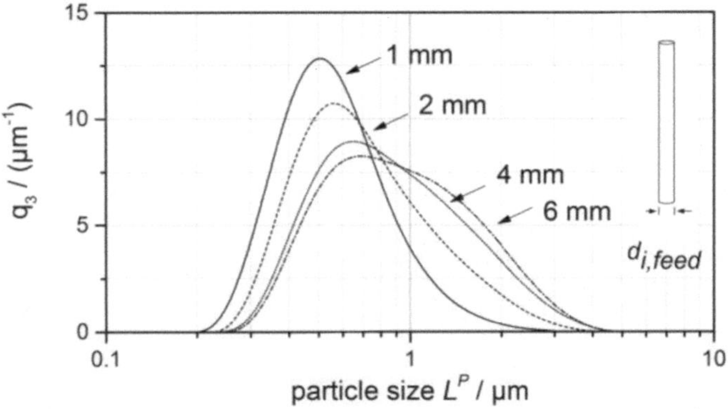

Figure 8.18: Influence of the feed pipe diameter on the PSD for $n = 100\ rpm$ *and* $\dot{V}_{\text{feed}} = 100\ ml/min$.

8.4.2 Reaction compartments R1

Figure 8.19 shows images of the two concepts realized for the reaction compartment R_1. The ST reactor (section 5.4.2) is on the left-hand side and the JiC configuration (section 0) is shown on the other side.

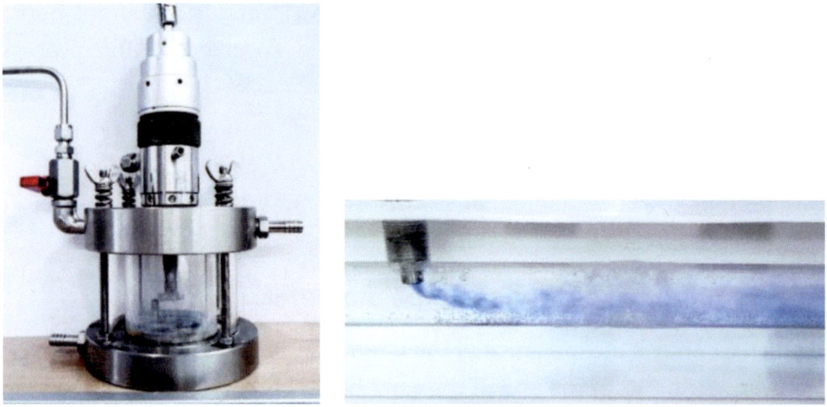

Figure 8.19: Pictures of the experimental reaction compartments R_1; left: ST reactor, right: JiC using ink for the flow visualization.

8.4.3 Turbulence model and sliding mesh procedure

The k-ε RNG model by Yakhot et al. [154] used to simulate turbulence in the ST reactor reads as eq. 8.14 and 8.15.

$$\frac{\partial}{\partial t}\rho k + \nabla \cdot \left(\rho k(\vec{u})\right) - \nabla \cdot \left[\left(\eta + \frac{\eta_t}{\sigma_k}\right)\nabla k\right] = P_k - \rho\varepsilon \tag{8.14}$$

$$\frac{\partial}{\partial t}\rho\varepsilon + \nabla \cdot \left(\rho\varepsilon(\vec{u})\right) - \nabla \cdot \left[\left(\eta + \frac{\eta_t}{\sigma_\varepsilon}\right)\nabla \varepsilon\right] = C_{1\varepsilon}\frac{\varepsilon}{k}P_k - C_{2\varepsilon}^{*}\rho\frac{\varepsilon^2}{k} \tag{8.15}$$

The constants and parameters used, such as $C_{1\varepsilon}$, $C_{2\varepsilon}^{*}$, P_k, can be extracted from the literature.

Transient ST reactor simulations are carried out by using the sliding mesh method. This method, applied to account for the rotating impeller, uses a modified formulation of the transport equation for a transported scalar ϕ. Thereby, a subtrahend $\vec{u}_r$ is added, which represents the motion speed of the rotating mesh.

$$\frac{\partial}{\partial t}\rho\phi + \nabla \cdot \left(\rho\phi(\vec{u} - \vec{u}_r)\right) - \nabla \cdot (D_t \nabla\phi) = \dot{S}_\phi \tag{8.16}$$

Figure 8.20 gives an idea on how the sliding mesh procedure works. The rotating (inner) cylinder moves with a certain predefined angular velocity, whereas the outer mesh is fixed. In the simulation executed, the correct functionality is increased by meshing the inner and outer interface area with the identical grid specifications.

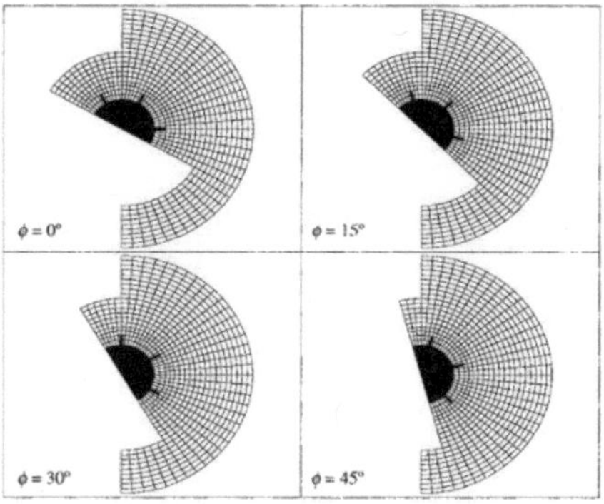

Figure 8.20: Sliding mesh procedure used for the CSTR simulated, cited from Ng et al. [155]

8.5 Mathematical annotations

8.5.1 DQMOM algorithm for CFD

This section presents a detailed description on how to implement the DQMOM algorithm into the CFD framework. Marchisio and Fox [16] and Fan et al. [17] are recommended literature. As described in section 2.1.2, the moments of a PSD can be represented mathematically consistent according to eq. 8.17.

$$\mathcal{M}_k = \int_{-\infty}^{+\infty} (L^P)^k f(L^P) dL = \sum_{i=1}^{N} w_i (\mathcal{L}_i)^k \qquad 8.17$$

For N=2, the first four moments distribution read as eq. 8.18.

$$\begin{aligned} \mathcal{M}_0 &= w_1 + w_2 \\ \mathcal{M}_1 &= w_1\mathcal{L}_1 + w_2\mathcal{L}_2 \\ \mathcal{M}_2 &= w_1{\mathcal{L}_1}^2 + w_2{\mathcal{L}_2}^2 \\ \mathcal{M}_3 &= w_1{\mathcal{L}_1}^3 + w_2{\mathcal{L}_2}^3 \end{aligned} \qquad 8.18$$

Equation 8.19 shows the DQMOM transport equations for the weights w_i and the weighted abscissa l_i $(= w_i\mathcal{L}_i)$ that must be solved within CFD neglecting the diffusion term. This combined solution of w_i and $\mathcal{L}_i$ ensures that the moments of the distribution stay conserved mathematically in the progress of the particle formation process.

$$\frac{\partial w_i}{\partial t} + \nabla \cdot (\vec{u} \cdot w_i) = a_i(w_i, \mathcal{L}_i)$$
$$\frac{\partial l_i}{\partial t} + \nabla \cdot (\vec{u} \cdot l_i) = b_i(w_i, \mathcal{L}_i) \qquad 8.19$$

According to Marchisio and Fox [16], the DQMOM approach determines the source terms a_i and b_i directly by the linear equation system in eq. 8.20.

$$(1-k)\sum_{i=1}^{N} \mathcal{L}_i{}^k\, a_i + k\sum_{i=1}^{N} \mathcal{L}_i{}^{k-1} b_i = \bar{S}_k \qquad 8.20$$

The linear equation system can be transposed into matrix style according to eq. 8.21 and 8.22 [17].

$$\mathcal{B}\vec{\theta} = \vec{\psi}$$
$$\mathcal{B} = [\mathcal{B}_1\mathcal{B}_2] \qquad 8.21$$

$$\mathcal{B}_1 = \begin{bmatrix} 1 & \cdots & 1 \\ 0 & & 0 \\ -\mathcal{L}_1{}^2 & & -\mathcal{L}_N{}^2 \\ \vdots & \ddots & \vdots \\ 2(1-N)\mathcal{L}_1{}^{2N-1} & \cdots & 2(1-N)\mathcal{L}_N{}^{2N-1} \end{bmatrix}$$

$$\mathcal{B}_2 = \begin{bmatrix} 0 & & 0 \\ 1 & \cdots & 1 \\ 2\mathcal{L}_1 & & 2\mathcal{L}_N \\ \vdots & \ddots & \vdots \\ 2(N-1)\mathcal{L}_1{}^{2N-2} & \cdots & 2(1-N)\mathcal{L}_N{}^{2N-2} \end{bmatrix} \qquad 8.22$$

$$\vec{\theta} = [a_1 \dots a_N\; b_1 \dots b_N]^T = \begin{bmatrix} a \\ b \end{bmatrix}$$

$$\vec{\psi} = [\bar{S}_0^N \dots \bar{S}_{2N-1}^N]$$

The source terms a and b can be found by matrix inversion of $\mathcal{B}$.

$$\vec{\theta} = \mathcal{B}^{-1}\vec{\psi} \qquad 8.23$$

The matrix $\mathcal{B}$ that must be inverted for $N = 2$ reads as eq. 8.25.

$$\mathcal{B} = \begin{bmatrix} 1 & 1 & 0 & 0 \\ 0 & 0 & 1 & 1 \\ -\mathcal{L}_1{}^2 & -\mathcal{L}_2{}^2 & 2\mathcal{L}_1 & 2\mathcal{L}_2 \\ -2\mathcal{L}_1{}^3 & -2\mathcal{L}_2{}^3 & 3\mathcal{L}_1{}^2 & 3\mathcal{L}_2{}^2 \end{bmatrix} \tag{8.24}$$

The inversion and the combination with eq. 8.23 leads to eq. 8.25.

$$\begin{bmatrix} a_1 \\ a_2 \\ b_1 \\ b_2 \end{bmatrix} = \begin{bmatrix} (3\mathcal{L}_1 - \mathcal{L}_2)L_2^2 & -6\mathcal{L}_1\mathcal{L}_2 & 3(\mathcal{L}_1 - \mathcal{L}_2) & -2 \\ (\mathcal{L}_1 - 3\mathcal{L}_2)L_1^2 & 6\mathcal{L}_1\mathcal{L}_2 & -3(\mathcal{L}_1 - \mathcal{L}_2) & +2 \\ 2\mathcal{L}_1^2\mathcal{L}_2^2 & -(4\mathcal{L}_1{}^2 + \mathcal{L}_1\mathcal{L}_2 + \mathcal{L}_2^2)L_2 & 2(\mathcal{L}_1{}^2 + \mathcal{L}_1\mathcal{L}_2 + \mathcal{L}_2^2) & -\mathcal{L}_1 - \mathcal{L}_2 \\ -2\mathcal{L}_1^2\mathcal{L}_2^2 & (\mathcal{L}_1{}^2 + \mathcal{L}_1\mathcal{L}_2 + 4\mathcal{L}_2^2)L_1 & 2(\mathcal{L}_1{}^2 + \mathcal{L}_1\mathcal{L}_2 + \mathcal{L}_2^2) & \mathcal{L}_1 + \mathcal{L}_2 \end{bmatrix} \cdot \frac{1}{(\mathcal{L}_1 - \mathcal{L}_2)^3} \cdot \begin{bmatrix} S_0 \\ S_1 \\ S_2 \\ S_3 \end{bmatrix} \tag{8.25}$$

The matrix that must be inverted for $N = 3$ is presented in eq. 8.26. The procedure leads to an extensive solution which should be solved preferably with computational methods. $N = 2$ is chosen for the work presented.

$$A = \begin{bmatrix} 1 & 1 & 1 & 0 & 0 & 0 \\ 0 & 0 & 0 & 1 & 1 & 1 \\ -\mathcal{L}_1{}^2 & -\mathcal{L}_2{}^2 & -\mathcal{L}_3{}^2 & 2\mathcal{L}_1 & 2\mathcal{L}_2 & 2\mathcal{L}_3 \\ -2\mathcal{L}_1{}^3 & -2\mathcal{L}_2{}^3 & -2\mathcal{L}_3{}^3 & 3\mathcal{L}_1{}^2 & 3\mathcal{L}_2{}^2 & 3\mathcal{L}_3{}^2 \\ -3\mathcal{L}_1{}^4 & -3\mathcal{L}_2{}^4 & -3\mathcal{L}_3{}^4 & 4\mathcal{L}_1{}^3 & 4\mathcal{L}_2{}^3 & 4\mathcal{L}_3{}^3 \\ -4\mathcal{L}_1{}^5 & -4\mathcal{L}_2{}^5 & -4\mathcal{L}_3{}^5 & 5\mathcal{L}_1{}^4 & 5\mathcal{L}_2{}^4 & 5\mathcal{L}_3{}^4 \end{bmatrix} \tag{8.26}$$

After converting the source terms into the form $\vec{\theta} = \mathcal{B}^{-1}\vec{\psi}$, the source term vector $\vec{\psi}$ for the specific moments k can be implemented with the following expressions. The source term for the particular moments can be calculated by the sum of mechanisms considered.

$$S_k = \text{Nucleation} + \text{Growth } (+ \text{ Aggregation} + \text{Agglomeration} + \text{Breakage})$$

Only nucleation and growth are considered in this work. Nucleation is implemented consonant with eq. 8.27.

$$S_{\text{Nuc},k} = B(x,t)(L_{\text{crit}}^{P})^k \tag{8.27}$$

In the case that the implementation of nucleation with a Dirac delta function at L_{crit}^{P} leads to instabilities, Hartig et al. [156] propose a formulation by a

Gaussian distributed critical particle size with a standard deviation σ^2. In the moment framework for the first four moments, this reads as:

$$\begin{aligned} S_{\mathrm{Nuc},0} &= B(x,t) \\ S_{\mathrm{Nuc},1} &= B(x,t) \cdot L_{\mathrm{crit}}^{P} \\ S_{\mathrm{Nuc},2} &= B(x,t) \cdot \left(\left(L_{\mathrm{crit}}^{P}\right)^2 + \sigma^2\right) \\ S_{\mathrm{Nuc},3} &= B(x,t) \cdot \left(\left(L_{\mathrm{crit}}^{P}\right)^3 + 3L_{\mathrm{crit}}^{P}\sigma^2\right) \end{aligned} \quad 8.28$$

Growth is implemented following eq. 8.29.

$$S_{G,k} = k \sum_{i=1}^{N} w_i \, \mathcal{L}_i^{\,k-1} G(\mathcal{L}_i) \quad 8.29$$

The source terms for aggregation, agglomeration or breakage are not considered (see section 2.4). If needed, they can be extracted from Fan et al. [17] and Marchisio and Fox [16].

The transport equations of the weights w_i and weighted abscissas ($l_i = w_i \mathcal{L}_i$) are linked to the specific Eulerian solid phase volume fraction α_i in the multifluid model to implement the DQMOM model into CFD [17]. (In our case the mixture model is used.) Thereby, comparable to section 2.3.2, sphere-like particles with $k_v = \pi/6$ are assumed.

$$\alpha_i = k_v w_i \mathcal{L}_i^3 = k_v \frac{l_i^3}{w_i^2} \quad 8.30$$

$$\alpha_i L_i = k_v w_i \mathcal{L}_i^4 = k_v \frac{l_i^4}{w_i^3} \quad 8.31$$

Thus, according to Fan et al. [17], the transport equations for the weights and weighted abscissas can be written as eq. 8.32 and 8.33.

$$\rho^S \left(\frac{\partial \alpha_i}{\partial t} + \nabla \cdot (\alpha_i \vec{u}) \right) = a_i' \quad 8.32$$

$$\rho^S \left(\frac{\partial \alpha_i \mathcal{L}_i}{\partial t} + \nabla \cdot (\alpha_i \mathcal{L}_i \vec{u}) \right) = b_i' \quad 8.33$$

The source terms a_i and b_i are also transcribed in this context (eq. 8.34).

$$\begin{bmatrix} a_1' \\ a_2' \\ b_1' \\ b_2' \end{bmatrix} = \begin{pmatrix} -2k_v\mathcal{L}_1^3\rho^S & 0 & 3k_v\mathcal{L}_1^2\rho^S & 0 \\ 0 & -2k_v\mathcal{L}_2^3\rho^S & 0 & 3k_v\mathcal{L}_2^2\rho^S \\ -3k_v\mathcal{L}_1^4\rho^S & 0 & 4k_v\mathcal{L}_1^3\rho^S & 0 \\ 0 & -3k_v\mathcal{L}_2^4\rho^S & 0 & 4k_v\mathcal{L}_2^3\rho^S \end{pmatrix} \cdot \begin{bmatrix} a_1 \\ a_2 \\ b_1 \\ b_2 \end{bmatrix} \qquad 8.34$$

a_i' represents the the source term of mass (eq. 8.32) and b_i' the source term of the abscissa $\mathcal{L}_i$ (eq. 8.33).

8.5.2 Product difference algorithm

The product difference algorithm [157] is used to calculate weights and abscissas from the moments of the distribution. This procedure represents one possibility to transfer a discrete, class-based distribution into weights and abscissas used within the DQMOM method. It represents the inverse step of eq. 8.17. Consonant with Gordon [157], the approach is built up basically following the subsequent procedure. A recursive matrix P is created initially. The entries of this matrix are specified as follows, wherein N represents the number of nodes (consisting out of an abscissa and a weight) that must be created. The first column is occupied by the Kronecker delta $\delta_{i,j}$ (eq. 8.35).

$$P_{i,1} = \delta_{i,1}$$
$$\delta_{i,j} = \{ \begin{matrix} 1 \text{ if } i = j \\ 0 \text{ if } i \neq j \end{matrix} \qquad 8.35$$

The entries for the second column read as eq. 8.36.

$$P_{i,1} = (-1)^{i-1} \cdot m_{i-1}$$
$$i \in 1, 2, \dots, 2 \cdot N + 1 \qquad 8.36$$

The other columns are completed with the product difference algorithm (eq. 8.37).

$$P_{i,j} = P_{1,j-1}P_{i+1,j-2} - P_{i,j-2}P_{i+1,j-1}$$
$$i \in 1, 2, \dots, 2 \cdot N + 2 - j \qquad 8.37$$
$$j \in 3, \dots, 2 \cdot N + 1$$

Thus, the matrix P with the moments $\mathcal{M}_{\mathrm{i}}$ for an $N = 2$ reads as eq. 8.38.

$$P = \begin{bmatrix} 1 & \mathcal{M}_0 & \mathcal{M}_1 & (\mathcal{M}_0\mathcal{M}_2 - {\mathcal{M}_1}^2) & \mathcal{M}_0(\mathcal{M}_3\mathcal{M}_1 - {\mathcal{M}_2}^2) \\ 0 & -\mathcal{M}_1 & -\mathcal{M}_2 & -(\mathcal{M}_0\mathcal{M}_3 + \mathcal{M}_2\mathcal{M}_1) & 0 \\ 0 & \mathcal{M}_2 & \mathcal{M}_3 & 0 & 0 \\ 0 & -\mathcal{M}_3 & 0 & 0 & 0 \\ 0 & 0 & 0 & 0 & 0 \end{bmatrix} \quad 8.38$$

Out of this matrix P, a further matrix Q is calculated, wherein the coefficients are determined as eq. 8.39. The procedure how to build matrix Q is well explained in Gordon [157] and should be extracted therefrom.

$$\alpha_i = \frac{P_{1,i+1}}{P_{1,i}P_{1,i-1}} \quad i \in 2, \dots, 2 \cdot N \quad 8.39$$

As explained by Gordon [157] the eigenvalues of matrix Q represent the abscissas $\mathcal{L}_i$. The weights can be calculated by the Eigenvectors (eq. 8.40). v_{j1} represents the first component of eigenvector j. A check of the result can be made by reinserting the weights and abscissas calculated in eq. 8.17.

$$w_j = m_0 v_{j1}^2 \quad 8.40$$

List of publications

Publications

1) M. Kind, P. Lau, **L. Metzger**, CFD-Rechnungen zur numerischen Messung von Zustandsgrößen in komplexen dynamischen Feststoffprozessen; *Chemie Ingenieur Technik*, 2016; doi: 10.1002/cite.201650298.

2) **L. Metzger,** M. Kind, The influence of mixing on fast precipitation processes – A coupled 3D CFD-PBE approach using the Direct Quadrature Method of Moments (DQMOM); *Chemical Engineering Science*, 2016; doi: 10.1016/j.ces.2016.07.006.

3) **L. Metzger**, M. Kind, On the mixing in confined impinging jet mixers – Time scale analysis and scale-up using CFD coarse-graining methods; *Chemical Engineering Research and Design*, 2016; doi: 10.1016/j.cherd.2016.02.019.

4) **L. Metzger,** M. Kind, Influence of mixing on particle formation of fast precipitation reactions – A new coarse graining method using CFD calculations as a "measuring" instrument; *Chemical Engineering Research and Design*, 2016, doi: 10.1016/j.cherd.2016.01.009.

5) **L. Metzger**, M. Kind (2015), The influence of mixing on fast precipitation processes – A coupled CFD-PBE approach using the Direct Quadrature Method of Moment (DQMOM); *Proceedings of the 11th International Conference on CFD in the Minerals and Process Industries*, 7 - 9.12. 2015, Melbourne, Australia, ISBN 978-1-4863-0620-6.

6) **L. Metzger**, M. Kind, Influence of mixing on particle formation of fast precipitation reactions – Coarse graining methods using CFD as a measurement instrument; *Proceedings of the 15th European Conference on Mixing, Mixing 15*, 28.06 - 03.07.2015, St. Petersburg, Russia, ISBN 978-5-905240-52-2.

7) L. Metzger, M. Kind, Supersaturation buildup in confined impinging jet mixers - A multi-scale coupling of CFD and population balance; *Proceedings of the 19th International Symposium on Industrial Crystallization, ISIC 19*, 16 - 19.09.2014, Toulouse, France.

8) **L. Metzger**, M. Kind, Compartment method for dynamic multi-scale simulation of precipitation reactors; *Proceedings of the 4th Joint US-European Fluids Engineering Division Summer Meeting*, 3 - 7.08.2014, Chicago, Illinois, Volume 1A, Applications in CFD, Paper No. FEDSM2014-21547, pp. V01AT03A013; doi: 10.1115/FEDSM2014-21547.

9) **L. Metzger**, M. Kind, On the transient flow characteristics in confined impinging jet mixers – CFD simulation and experimental validation, *Chemical Engineering Science*, 2015, 133, 91-105. doi: 10.1016/j.ces.2014.12.056

Conference contributions

1) **L. Metzger**, M. Kind, Dynamische Simulation technischer Fällprozesse (Talk), *ProcessNet Fachausschuss Kristallisation*, 10 - 11.03.2016, Frankfurt am Main
2) **L. Metzger**, M. Kind, The influence of mixing on fast precipitation processes – A coupled CFD-PBE approach using the Direct Quadrature Method of Moment (DQMOM) (Talk), *11th International Conference on CFD in the Minerals and Process Industries*, 7 - 9.12.2015, Melbourne, Australia
3) **L. Metzger**, M. Kind, Influence of mixing on particle formation of fast precipitation reactions – Coarse graining methods using CFD as a measurement instrument (Talk), *Proceedings of the 15th European Conference on Mixing, Mixing 15*, 28.06 - 03.07.2015, St. Petersburg, Russia
4) **L. Metzger**, M. Kind, Einfluss der Vermischung auf die Partikelbildung von schnellen Fällungsreaktionen – Untersuchung mit CFD basierten Methoden (Talk), *ProcessNet Fachausschuss Kristallisation*, 18 - 20.03.2015, Magdeburg
5) **L. Metzger**, M. Kind, Supersaturation buildup in confined impinging jet mixers - A multi-scale coupling of CFD and population balance (Talk), *19th International Symposium on Industrial Crystallization*, ISIC 19, 16 - 19.09.2014, Toulouse, France
6) **L. Metzger**, M. Kind, Compartment method for dynamic multi-scale simulation of precipitation reactors (Talk), *4th Joint US-European Fluids Engineering Division Summer Meeting*, 3 - 7.08.2014, Chicago, Illinois
7) **L. Metzger**, M. Kind, Vernetzte Methoden zur dynamischen Simulation von Fällungsreaktoren (Talk), *ProcessNet-Jahrestreffen der Fachgruppe Kristallisation*, 27 - 28.3.2014, Münster
8) **L. Metzger**, M. Kind, Vernetzte Methoden zur dynamischen Simulation von Fällungsreaktoren (Poster), *ProcessNet Fachausschuss Computational Fluid Dynamics*, Mischvorgänge und Rheologie, 24 – 26.2.2014, Würzburg
9) **L. Metzger**, M. Kind, CFD modeling of precipitation processes (Poster), *PARTEC 2013 – International Congress on Particle Technology*, 23 – 25.4.2013, Nürnberg
10) **L. Metzger**, K. Nguyen, P. Hirschel, M. Kind, CFD-Simulation von Fällungsprozessen (Poster), *ProcessNet Fachausschuss Kristallisation*, 14 - 15.03.2013, Magdeburg
11) **L. Metzger**, J. Udri, R. Kügler, M. Kind, CFD modeling of precipitation processes (Poster), Nano-structure in disperse systems: production, characterization and functionality, 12 - 13.11.2012, Karlsruhe

12) R. Kügler, J. Morgenstern, **L. Metzger**, M. Kucher, M. Kind, Fällungskristallisation von Bariumsulfat - Vermischung der Edukte und Stabilisierung von Nanosuspensionen (Poster),
ProcessNet Fachausschuss Kristallisation, 5 - 6.03.2012, Wittenberg

Student theses conducted in conjunction with this thesis

1) **Michael Zeller**, CFD Simulation von Rührkesseln – Bestimmung prozessbestimmender Parameter für die Fließschemasimulation, Bachelor.

2) **Jan Sauer**, Experimentelle Untersuchung technischer Fällungsreaktoren, Bachelor.

3) **Tobias Nachtrab**, Aufbau und Inbetriebnahme einer Kompartmentanlage für die experimentelle Simulation technischer Fällapparate, Master.

4) **Marijan Mohr**, Fließschemasimulation von technischen Fällungsreaktoren, Bachelor.

5) **Sonja Seiler**, Untersuchung prozessbestimmender Zonen im Rührkessel mit CFD-Methoden, Bachelor.

6) **Matthias Kunz**, Modellentwicklung zur Berechnung von technischen Fällungsreaktoren, Master.

7) **Marvin Winkler**, Messung von Verweilzeitverhalten und Übersättigungsaufbau in statischen Mischdüsen in verschiedenen Strömungsregimen, Bachelor.

8) **David Guse**, Untersuchung einer modifizierten Mischdüse für die Fällungskristallisation, Bachelor.

9) **Kim Nguyen**, Numerische Untersuchung und Gitteroptimierung von statischen Mischdüsen, Study work.

10) **Klaus Müller**, Charakterisierung der Interaktion von partikulärer Phase und fluider Phase bei der Fällungskristallisation von Bariumsulfat, Bachelor.

11) **Pascal Hirschel**, Untersuchung der Vermischung in Mischdüsen bei der Fällung mit einer Spritzenpumpe, Bachelor.

12) **Jan Udri**, Untersuchung der Strömungsmechanik in Mischdüsen mittels Particle Image Velocimetry (PIV)-Messungen, Bachelor